TECHNOLOGY TODAY

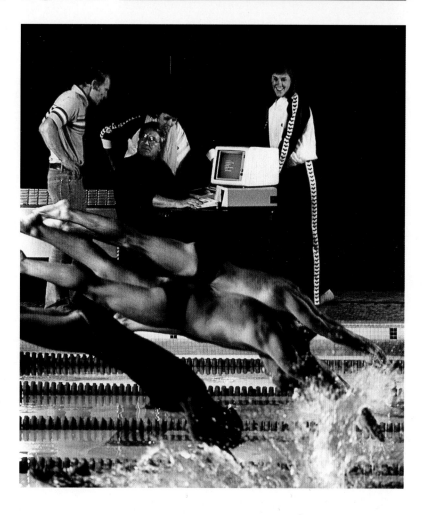

Technology is developed by humans. Humans are changed by technology. Technical advances have caused changes in our world since people developed stone tools millions of years ago.

The computer has caused great changes in modern culture. People think and act differently today as a result of computers. This modern human-made tool is an example of the effect technology has on people.

Computers are used in nearly every human activity. Race times for swimmers are recorded more accurately than can be done by people with stop watches. An electronic starter's "gun" signals the start of the race and starts the clock. The race times of the swimmers are recorded when they touch a pad in the water at the end of the pool. (Courtesy IBM)

PROCESSING

This is an automated processing plant. It was sent to Mars where it took samples of the planet's surface with the scoop on the long arm. The samples were dumped into a hopper in the body of the craft. The samples were then tested to determine their composition. This information was then transmitted to scientists on earth. (Courtesy of Martin Marietta)

A modern oil refinery looks like a huge metal sculpture. Actually it is a marvel of modern engineering where hundreds of chemical processes occur under closely controlled conditions.

Computers are used to monitor and control the operations in the processing of crude oil. The results are the familiar petroleum products such as gasoline, waxes, greases and materials for making plastics. (Courtesy of Kerr-McGee Corp.)

Steel beams are essential to modern construction. White-hot steel is processed into beams at a rolling mill. When heated to very high temperature, long strips of steel can be shaped into the desired form by rollers. The steel tracks for railroads are produced in this same way. This method, called <u>forming</u>, changes the shape of steel while increasing its structural strength. Large amounts of heat energy are necessary to perform this process. (Courtesy of the American Iron and Steel Institute)

This large tub holds 4,500 kilograms (10,000 lbs.) of chocolate paste. The metal arms in the tank mix the chocolate. Here, the technician is preparing to pump the chocolate mixture to another part of the processing plant. This room shows just one stage in the process of converting cacao beans (the raw material) into chocolate bars, chocolate syrup or cocoa (finished products). Many foods are processed from raw form into finished products. For example, wheat is ground into flour which is then mixed with yeast and water and baked to make bread. What are some other foods that need processing? (Courtesy of Hershey Food Corp.)

ENERGY

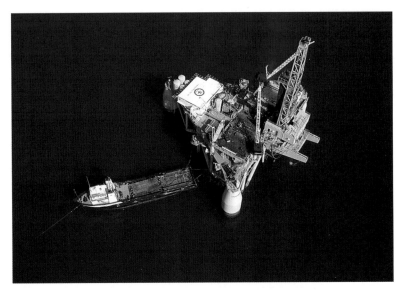

This self-contained platform is designed to support one or more drilling rigs. Drilling supplies, production equipment and living space for 50 or 60 workers are also on the platform.

The semi-submersible drilling platform is towed to its position in the ocean. Water is drawn into the three corner cylinders to make the structure submerge enough so that wave motions have little effect on it. (Courtesy of Kerr-McGee Corp.)

The process of producing oil requires coordination between men and machines. Sea-going oil men work at the well head. A string of drill pipe is lowered into the casing and is turned. The drill bit on the end of the string cuts into the ocean floor. As the drill pipe moves down more sections of pipe must be added.

Periodically the long string is raised one section at a time to attach a new bit. When the new bit is attached the string of pipe is reassembled, one piece at a time, and lowered back into the well. Drilling continues around the clock for weeks or months before the undersea oil is reached. (Courtesy of Kerr-McGee Corp.)

For years researchers have tried to develop a practical electric-powered automobile. The electric car offers the promise of eliminating the air pollution caused by internal combustion engines. Electric powered vehicles are used in warehouses because they do not pollute the air and run quietly.

Batteries needed for electrical energy still take up large amounts of space for the amount of energy produced. The batteries also must be recharged frequently limiting the length of trip possible without stopping to recharge. (Courtesy of General Electric)

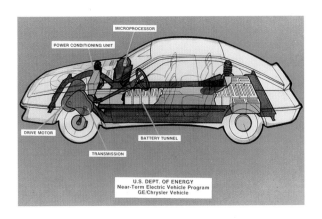

The Venus Radar Mapper (VRM) Mission will place a spacecraft into orbit around Venus to map its surface. The mapping will allow scientists to study the geology of the planet. Because of the heavy cloud cover on Venus, microwave energy will be used for the mapping instead of conventional photography. Pictures taken of the surface of Venus will be transmitted back to scientists of earth.

The two "wings" on the spacecraft contain thousands of solar cells. These provide the energy needed to run the many electrical systems on the craft. The cells convert radiant energy from the sun into electrical energy. The electrical energy can be used directly or be used to recharge batteries. By using the readily available solar energy, the spacecraft can remain in service for a long time without returning to earth. (Courtesy of Martin Marietta)

MANUFACTURING

Robots have many applications in manufacturing. Here they are being used to weld truck bodies. They can do jobs that would be dangerous for humans. Precision welding can be done without exposing workers to the heat, fumes and flying sparks. The robot welders also can do work faster than humans. Accuracy can be improved resulting in a more reliable, manufactured product.

The use of automated systems including robots in manufacturing has resulted in the loss of jobs for the workers. Robots can do the work of several workers. What effect has this had on the number of people who are out of work? While more workers are needed to build and repair the robots, these are more highly skilled workers than the ones being displaced. (Photo courtesy Ford Motor Corp.)

Intricate shapes can be cut in metal using intense laser beams. A machine part can be made very accurately with a minimum of wasted material. Lasers are also used in very delicate eye surgery. The laser beam can be focused on a very small spot inside the eye that could not be reached by other methods of surgery. These are examples of similar technologies used in different ways. (Courtesy of Spectra-Physics)

The latest in technological advances are used to design and manufacture even commonplace objects such as clocks. Modern electronics were used in the quartz crystal control circuit of this clock. Stepping motors and drive mechanisms were added. The result was a clock that was superior in its design and performance. Modern manufacturing techniques were applied resulting in a successfully marketed product. (Courtesy of Arthur D. Little, Inc.)

Some manufacturing operations must be done in a very clean and controlled environment. Even small specks of dust or a loose human hair could ruin a part when it is being manufactured. Contrast this "clean room" environment with the work area shown on the opposite page. Welding or spraying of automobile bodies are done in less clean areas. While cleanliness is important, a "clean room" is not necessary. (Courtesy of IBM)

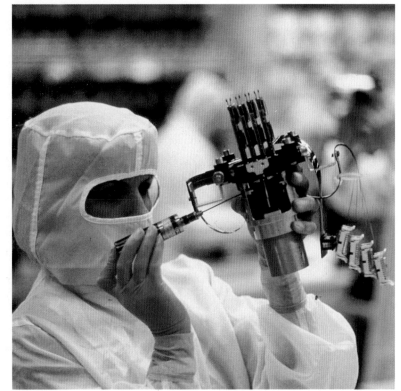

CONSTRUCTION

Space stations now being designed will be constructed with a "dual keel design" using booms to connect several modules. Sky-viewing instruments will be placed on the top boom, while earth-viewing instruments will be mounted on the lower boom. Each boom will be constructed of trusswork similar to the trusses used on earthbound structures. Can you find trusses that look similar to the designs used on the space station booms? Look at bridges, even in the structure of buildings. (Courtesy of Boeing Aerospace, above; and NASA, below).

The materials used in construction have changed over the years. The use of steel and glass in modern high-rise buildings contrasts with the stone commonly used just a short time ago. Technological advances in construction have changed the skyline of most cities.

Modern materials have made newer construction methods possible. Before the development of modern, high-strength steels there was a limit to the height a structure could be built. Modern structural glass can be used as an outside covering for buildings where brick and stone were used. (Courtesy of Arthur D. Little, Inc.)

TRANSPORTATION

Rockets are also transportation vehicles. They carry spacecraft into space above the earth. These spacecraft can be manned, such as the space shuttle, or unmanned, such as weather and communication satellites. Rockets must be powerful enough to escape the gravitational pull of earth. This rocket is carrying into space the INTELSAT communications satellite pictured on p. xiii. (Courtesy of General Dynamics)

Modern transportation vehicles take many forms. The medium through which the vehicle moves dictates the shape of the vehicle. Ships have to be made so they will float, carry large loads and move through water. Some small ships can rise above the water on wing-like devices called hydrofoils. Large ships look much the same above the water as they did 50 years ago. But significant design changes in the hull below the water have greatly increased their efficiency. (Courtesy of General Dynamics, left; and Boeing Marine Systems, right)

Conveyors are an important part of in-plant manufacturing systems. Auto bodies travel on a trolley which straddles a power track. Conveyors can be mounted on the floor or overhead using ceiling supports. Workers can add parts, make adjustments, even provide final touches to the paint work. When work must be done on the underside of an auto body, the conveyor system can be designed to carry the auto body above the workers. (Courtesy of Jervis B. Webb Co.)

COMMUNICATION

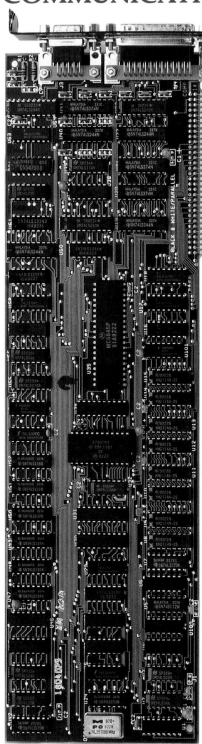

The computer circuit board is a marvel of electronic miniaturization. These boards can do the work of a room full of circuits using vacuum tubes. The integrated circuit chips on these boards perform functions many times faster, run cooler and require very little energy to operate.

Many commonly used machines in the home rely on integrated circuits, most much smaller than this. The electric range, clothes dryer, television, video recorder, and even the family car may be at least partially controlled by computer circuitry. (Courtesy of IBM).

Drafting and design have changed dramatically in only a few years. Computer aided drafting (CAD) has opened new worlds to the designer. This computer screen shows a three-dimensional geometric model. The computer makes it possible to visualize the object from many positions. The object can be moved around on the computer screen. Individual parts can be "zoomed-in" so small details can be seen. Multicolor graphics add greatly to the designer's ability to visualize an object. (Courtesy of Arthur D. Little, Inc.)

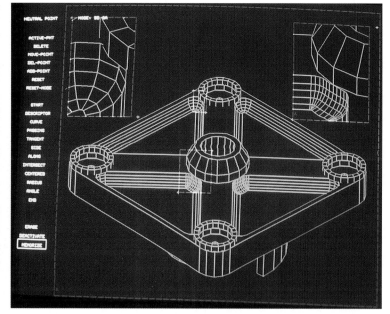

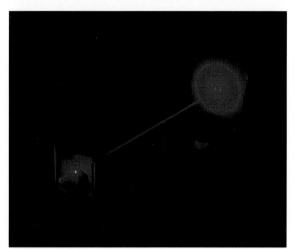

Laser technology is being applied in many fields such as electronic packaging and business communication systems. Very accurate testing, measurement and quality control systems are possible because of the precise nature of the laser. Laser technology is used in computer aided design and engineering, signal processing and many areas of electronic communication. Lasers are used by land surveyors. In recent years the laser has been developed as a valuable tool in medicine, particularly in surgery of the eye and skin cancer. (Courtesy of Arthur D. Little, Inc.)

Satellite ground terminals send and receive signals to and from earth orbiting satellites. These highly sophisticated antennas were designed for Defense Department communications systems. The huge antenna, 18 meters (60ft) wide, can pick up very faint signals from space. They also can send signals to satellites orbiting the earth. Mechanisms are built into the antennas to allow them to be rotated and tilted. They can be aimed toward different satellites orbiting the earth. (Courtesy of Ford Aerospace)

Have you ever talked on the telephone to a person in another country? If so, you probably noticed a delay between the end of your message and the beginning of the other person's reply. This is caused by the time needed for the electronic signal to bounce off the communications satellite.

The INTELSAT advanced worldwide communications satellite is more than 50 feet tip-to-tip with the solar panel wings deployed and 22 feet high. It weighs more than 4,000 pounds. The hundreds of solar panels convert solar energy to electrical energy. This satellite can provide 15,000 two-way voice channels at the same time. (Courtesy of Ford Aerospace)

AUTOMATION

This copper manakin is an example of an automated system. It is used to test the insulating protection of thermal clothing. This system makes it unnecessary to subject humans to the harsh conditions of arctic cold and desert heat in the tests of new protective clothing. Built-in heaters simulate human body heat. Sensors are used in an automatic temperature control system.

Hip, knee, shoulder and elbow joints allow the manakin to "walk." The stride and arm swing can be adjusted. The manakin can simulate walking speeds from one stroke in two seconds to 1.2 strokes per second. (Courtesy of Arthur D. Little, Inc.)

Satellites provide automated information gathering. They can be programmed to photograph automatically specific areas of the earth's surface. This infrared photograph of the western end of Lake Erie was taken by NASA's Landsat satellite from a height of 912 km (570 miles). Infrared film shows vegetation as red. Water pollution from the urban areas of Toledo and Detroit show up as lighter areas on the shoreline. (Courtesy of NASA, USGS EROS Data Center)

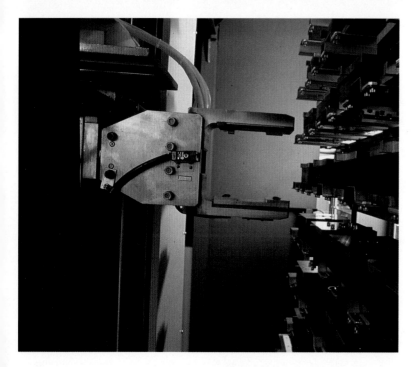

Robots can be used to perform delicate tasks. Fast and accurate assembly of electronic devices is possible with automated robotic equipment. The gripper on the end of a robot arm compares with the human hand and wrist. The gripper can be opened and closed, the "wrist" joint can be turned much as a human wrist is turned.

End-of-arm tooling (often called the end effector) can vary as needed for the work to be done. A welder, spray gun or wrench can be used instead of the hand-like gripper. Special purpose tooling can be used to perform many different jobs. (Courtesy of IBM)

Robots can be used in areas where humans would be exposed to hazardous work conditions. Here, a robot arm is used to spray-paint the chassis of a tractor. If humans were doing the spraying, they would be exposed to the toxic paint fumes and the danger of explosion.

The robot can be programmed to apply the proper amount of paint in the proper places. It can reach over the top of the assembly, around the ends and even underneath it if needed. When the assembly line moves another chassis into the position, the robot repeats the cycle of spraying. (Courtesy of Deere & Company)

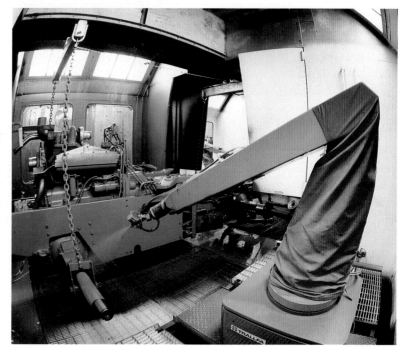

TECHNOLOGY: THE FUTURE

Can you imagine people living and working in a space station in this century? This is not a science fiction story. Work is under way to construct an orbiting space station. Living and working space will be in pressurized modules connected by booms. Up to five modules will be used to provide living quarters for the astronauts. Hanger space will be used to service spacecraft. Laboratories will be used for experiments such as zero-gravity processing of materials.

    The space station can be used as a "base camp" for astronauts who service, rescue or relaunch satellites. More space stations and platforms can be constructed from this base. Hangars and garages can be provided for housing and repairing mission vehicles.

    This space station concept is the starting point for a new era of benefit and discovery. Will it ever happen? As technology develops, you may see space stations such as this as a reality rather than merely a science fiction story. (Courtesy of Boeing Aerospace Co.)

Exploring Technology

Snowmaking equipment is an example of production technology. The ability to manufacture such artificial snow gives ski slope managers some control over the natural environment.

E. Allen Bame
Paul Cummings

Paul W. DeVore
Consulting Editor

Exploring Technology

SECOND EDITION

Delmar Publishers, Inc.
Albany, New York

NOTICE TO THE READER

Publisher does not warrant or guarantee any of the products described herein or perform any independent analysis in connection with any of the product information contained herein. Publisher does not assume, and expressly disclaims, any obligation to obtain and include information other than that provided to it by the manufacturer.

The reader is expressly warned to consider and adopt all safety precautions that might be indicated by the activities described herein and to avoid all potential hazards. By following the instructions contained herein, the reader willingly assumes all risks in connection with such instructions.

The publisher makes no representation or warranties of any kind, including but not limited to, the warranties of fitness for particular purpose or merchantability, nor are any such representations implied with respect to the material set forth herein, and the publisher takes no responsibility with respect to such material. The publisher shall not be liable for any special, consequential or exemplary damages resulting, in whole or in part, from the readers' use of, or reliance upon, this material.

Copyright 1988
Delmar Publishers, Inc.

For information, address Delmar Publishers
2 Computer Drive West, Box 15-015
Albany, New York 12212

All rights reserved. No part of this publication may be reproduced or transmitted in any form or by any means, electronic or mechanical, including photocopying, recording, or any storage and retrieval system now known or to be invented, except by a reviewer who wishes to quote brief passages in connection with a review written for inclusion in a magazine, newspaper, or broadcast.

Printed in the United States of America
Published simultaneously in Canada
by Nelson Canada,
A division of The Thomson Corporation

Library of Congress Catalog Card Number:
ISBN: 0-87192-196-0

10 9 8 7 6 5 4 3

Contents

Preface

Introduction

Chapter 1 What is Technology?   1
  Introduction   1
  How Things Have Changed   3
  How Does Technology Affect Us?   9
  Suggestions for Further Study   21
  Activity: Developing a Time Line   24

Chapter 2 Processing Technology   25
  Introduction   25
  Preprocessing   28
  Working in Processing   50
  Suggestions for Further Study   52
  Activity: Recycling Aluminum   54

Chapter 3 Energy and Power   57
  Introduction   57
  Energy Sources   60
  Energy Conversion   68
  Transmitting Energy and Power   88
  Using Energy   15
  Suggestions for Further Study   105
  Activity: Testing Efficiency   108

Chapter 4 Manufacturing   111
  Introduction   111
  Resources of Manufacturing   112
  Product Development   117
  Producing a Product   122
  Marketing   131
  Working in Manufacturing   134
  You—The Consumer of Manufactured Goods   138
  Suggestions for Further Study   140
  Activity: Pen Assembly Line   142

Chapter 5 Construction   145
  Introduction   145
  Dwellings   147
  Transportation   148
  Impact of Construction   153
  Technical Advances in Construction   154
  Construction to Meet Human Needs   160

Working in Construction   173
Suggestions for Further Study   176
Activity: Building and Testing Trusses   178

Chapter 6 Transportation   181
Introduction   181
Historical Development
  Before the Twentieth Century   182
Modern Transportation
  by Air and Conveyor Systems   186
Components of Transportation   197
Working in Transportation   232
Suggestions for Further Study   234
Activity: Designing an Egg Passenger Vehicle   236

Chapter 7 Communication   239
Introduction   239
History of Communication   241
Development of Telecommunication Media   249
Impact of Communication   269
Suggestions for Further Study   270
Activity: Producing a Radio Script   272

Chapter 8 Automation   275
Introduction   275
Components of Automation   275
Automation in Practice   280
Suggestions for Further Study   292
Activity: Robots   294

Chapter 9 Computers in Technology   297
Introduction   297
Developments of Computers   297
What is a Computer?   306
Computers Today   321
Suggestions for Further Study   322
Activity: Computer Programming   324

Chapter 10 Consequences of Technology   327
Introduction   327
Eight Counterforces of Technology   328
Systems of Technology   334
Suggestions for Further Study   345
Activity: Consequences   346

Glossary   347
Readings   355
Index   361
Picture Credits   365

Preface

Technology, once accepted without question, now greatly concerns us all. Most of us have placed great faith in our technical systems only to find that they often create harmful conditions for all forms of life.

We have become more aware of the impact of technical means on the nature of society and the quality of life during the twentieth century. In earlier times, humans had to know about and understand the natural environment in order to survive. It is now necessary to be aware of and understand the technological environment and its relationships among the human, social, and natural systems as well.

In past times, decisions made by individuals and groups about technology affected only the immediate family, social group, or local environment. Technology's power was limited. Today, however, these decisions can affect people throughout the world, often in destructive and unplanned ways.

Those who use this textbook will be living in the twenty-first century. The kind of life you live will depend on the decisions you make today and tomorrow about the nature of our technological systems. These decisions will depend on how well *all* citizens understand the technological, sociological, and ecological systems and their relations to human life.

Creating a society in which technology serves human needs will require new knowledge and understanding on the part of citizens. This series of publications was designed to serve a critical educational need that exists at all learning levels.

This book is one of three for the middle school level. The main purpose of each of these books is to provide an organized study of technology beginning with its creation, moving through the development of industry to present technical systems, including automation and cybernetics. This series is designed to support a program for the study of technology as part of basic education, kindergarten through college.

The goal of this book is to help us better manage our technical means. If we can do this, we will enter the twenty-first century by choice, not chance.

Morgantown, West Virginia
October 1987

Paul W. DeVore
Consulting Editor

Introduction

Exploring Technology is the second of a three-part series of publications for use in the study of technology in the middle school. The first part of the series deals with early mankind and the development of technology. This second level explores modern technological systems. The third level introduces the various elements, forms, and consequences of technology. In all three levels a historic perspective is developed to contribute to the complete understanding of modern systems. Projecting to the future can help see what is likely to happen as a result of current technology.

This book is about technology. This is very simple to say. But, as you will see, it can become complicated. So, *Exploring Technology* and the materials that accompany the text attempt to explain technology in simple, everyday ways. A definition of technology is given in chapter 1. You are encouraged to experience technology.

Like many other complicated things, technology is best learned a bit at a time. To do this it is helpful to learn the smaller details—the little bits and pieces. When you look at a picture for the first time you get a general impression. The longer you look at the picture the more detail you see. We have written this book to help you get the "big picture." As you continue to study technology you will learn more and more of the fine details.

Exploring Technology is intended to give you a good foundation on which to build further study. This can be compared to a pyramid—a "pyramid of learning." When the pyramids of Egypt were built, a wide and strong base was built first. Then the rest of the construction was put on top of that. It is important to build this strong base for further study.

You will be introduced to many concepts of technology. Concepts can be called "big ideas," the central points around which the finer points are built. The first chapter answers the question, "What is technology?" It attempts to show the relationships of technology to all of human life. After this introduction you will study the three major systems of technology—production, communication, and transportation—through several chapters in the book.

Processing, a part of the production system, is considered first. Through processing, the many materials we use are obtained from the earth and converted into useful form.

Energy and power are considered next. These important elements are common and necessary to all technological systems. Without energy and power little would be possible in technology.

Manufacturing and construction, also parts of production, transportation, and communication are considered next. These make up the major systems of technology. They are complex, but their understanding is critical to the whole of technology.

Chapters 8 and 9 look at some aspects of modern technology. Automation, the human-like working of systems under the command of machines, is the controller of modern technology. Computers are involved with every aspect of technology. They are also a part of all our lives.

The final chapter takes a brief look at the consequences of technology. People have benefited from technology in many ways. We also have suffered. The consequences are both positive and negative.

At the end of each chapter in this book there is a section called "Observations." Here you will find suggestions of ways you can check your understanding of the content of the chapter. Questions related to the reading in the chapter will be included. There will be a list of key words from the chapter. Some of these words may be new to you. If you have read the chapter carefully you should know their meaning. If not, look them up in a dictionary or encyclopedia. A glossary is included in the back of the book to help you in learning the meanings of the words.

The best way to learn about something is to become involved with it. Reading is fine, but actually doing it adds more to your learning. Several short activities are given at the end of each chapter. These are designed to get you involved in learning about technology. Many of these are written so you can do them at home. Some help you become involved in your community. Most can be done in a short time.

The activity manual that accompanies this book gives more activities you can do. These are usually larger and more detailed. Most of these activities are designed to be done in school. Your teacher may be using this manual in your class.

All technological systems are interrelated. None can exist without the others. Study them in this way. Move back and forth from chapter to chapter. Transportation vehicles need power plants. Power plants need fuel. Raw materials must be taken from the earth and processed to make fuel. Processing requires transportation vehicles and energy and power. The interrelatedness of technological systems becomes obvious once you begin the study by focusing on the total system and the relationships of the various parts.

Modern production systems help make available a wide variety of medicines.

Chapter 1   What is Technology?

Introduction

What is technology? What does the word *technology* mean to you? What images come to mind when you see or hear the word? The chances are, no image comes to mind or you think of some tool or machine. When you see the word *apple* you get a mental picture or image of a well-known object. But this is not so for technology. There is no image because technology is not a single object like an apple. No image forms also, possibly, because the word is new to you.

    Technology is very complex. It means different things to different people. A dictionary definition might read, "applied science" or, "all the ways used to provide those things necessary for human comfort." Another definition might be, "the ways humans use tools and machines to make their life easier." Whatever way it is defined, several points are important. (1) The study of technology deals with tools, machines, materials, techniques, and technical systems and the relationship of these elements to human beings and society. (2) Without people to develop it, there would be no technology.

    The earliest technological advancement probably occurred when a person in the stone age shaped a stone to use as a tool. Soon the stone was tied to a stick and a club was made. These were the first tools made by people. With the first ax or the club it was possible to make things. Animals could be killed for food. These simple tools made life a little easier for the humans of that time. It was technology in its simplest form.

    It is not as simple today. Examples of our technology are all around us. Look around you. What do you see that is a result of technology: the desk or chair you are sitting in, the electric light that helps you see this book? This book is a product of technology. It might be more difficult to look for things around you that are *not* examples of technology. There are far fewer of them. You will have to be careful in this search though. You could be fooled by technology.

Certainly the things in nature are not products of technology! But, are they? What about those pine trees on the hillsides? They may be "super trees," which have been developed by foresters to grow faster and to produce more wood pulp or lumber in a shorter period of time. The shrubs around your house or school may have been developed specially. Their growth is enhanced by fertilizer or by insecticides that are sprayed on them to keep off harmful insects.

What about the birds in the sky? Are they products of technology? They may not be products of technology but they are affected by it. Some species of birds are almost extinct because of poisons in their food. DDT (dichloro-diphynl-trichloroethane), for example, was found to be causing a decline in the numbers of many species of birds. The DDT was causing the birds to lay eggs with thin shells. These eggs were easily broken before they could hatch.

Technology surrounds us. It affects nearly every part of our lives. Mankind has developed to modern levels through technology. We can never go back to the ages before there was any technology.

Our use of technology results in many things. One of the most important is change. Technological advances always result in some form of change. The stone ax changed the life of humans in prehistoric times. The addition of stirrups to saddles made it much easier to ride a horse. As a result, transportation on horseback was easier and faster. Warfare changed because soldiers could ride faster and stay on their horses even in the middle of a hard battle. The development of mass-production techniques made the automobile less expensive. More people could afford to own an automobile. This resulted in the need for safer and better roads.

Technological change brings about changes in the way we do things. Our lives are changed along with our ways of thinking. Why do we change the way we produce, transport, or communicate things? We change because we want to work less and have more leisure time. We want better food to eat. We want to live in better and more comfortable houses. We want to go more places,

The Kodak #2 Brownie box camera was a popular camera in the early 1900's. Compare this camera with a modern 35mm camera. Notice how the design has changed. Technological progress can be seen in these products.

How Things Have Changed

with less effort, and in less time. We want better health and better care when we are ill.

A look at several areas of technical progress can help us see more of the changes. Remember, when we study this progress we are also studying the way mankind changes. Progress and change go together. They cannot be separated. One way to look at this progress is to look at the changes that have occurred over a period of time. As we compare the past with the present, the changes can be seen. This look at the past and present should also give some hints about the future. We must be aware of the fact that both positive and negative changes occur. We must look for both or we will have an inaccurate picture.

Resources and Materials

Technology has provided us with many materials that help make our lives better. Through technical progress we have learned to make materials into something useful. Before the development of modern steels and other metals people had only natural materials to use. Weapons had to be made from sticks, stones, and plant fibers. Before the development of drilling techniques, wood, coal, and whale oil were the main fuels used for light and heat. With the development of modern drilling techniques, large amounts of oil

Petroleum deposits under the ocean floor can now be tapped from an oil rig such as this one.

were available to be made into gasoline, kerosene, fuel oil, and thousands of other everyday products.

Modern materials such as plastics have made living better. Plastics have replaced wood, metals, ceramics, and textiles in many uses. Look at the materials we use every day. Most of them are products of technology developed in recent years. Celluloid, the first American plastic, was made just over one hundred years ago. Synthetic rubber was developed in the 1940s to offset shortages of natural rubber caused by World War II. Today synthetic rubber is used to make millions of automobile tires a year.

So, we can see that as a result of technology, people are able to develop new materials. Through technology we are able to make the materials of nature into something more useful to us. We are able to make new materials that do not exist in nature. Materials can be mixed or separated into smaller parts to produce new materials. They can be made harder, softer, thinner, thicker, fire resistant, or more flammable.

Tools and Materials

The changes made in tools and machines are obvious. These can be seen in any museum or display of old tools. Many people have private collections of old hand tools. These are good examples of technological change. When one begins to study old tools one thing becomes obvious. Early tools required a lot of muscle power by the craftsman. Modern tools have eliminated much of this. Over the years we have been able to reduce the labor associated with work. This is an important contribution of technology.

Tools and machines have affected our lives in many ways. Compare the modern home with homes of colonial America. The colonists made their homes entirely with hand tools. Trees were cut into logs, logs cut into lumber, and the lumber erected into the home. Human energy and hand tools such as the adz, plane, ax, and saw were mainly used. Today the trees are cut with a chain saw. Trucks haul the logs to the saw mill. Logs are cut into lumber by huge saws driven by diesel engines. Carpenters use power saws to cut the lumber. A back hoe mounted on a tractor is used to dig the foundation for the house. Transit trucks mix and deliver the concrete for the foundation.

Consider the impact of the computer on our modern world. Computers are used in automobiles to control the running of the engine. Computers control airplanes, spacecraft, ships and trains. You may be using a computer in your home or at school to write papers or to calculate math problems. This book was written on a computer. Computers are our newest machines. They help us do things easier, faster, and better.

The human being is the only creature on the earth to develop a technology. Though some animals do make tools, they have no advanced technology. Thus, humans are separated technologically from the other creatures on earth.

Energy and Power Systems

The growth of technology depends on energy and power. Without energy and power we could not have many of the things we enjoy. Until the early nineteenth century most work was done using muscle power. Human muscles and the muscles of work animals were the major sources of energy. Although other sources existed before that time such as water wheels, windmills, and some of the early steam engines, these sources provided only a small part of the total energy used.

Everything changed after the invention and development of the steam engine. The steam engine was used to provide the power to run factories and drive automobiles, ships, and trains. With the steam engine to provide power, factories became larger. Larger factories needed large numbers of workers. Towns and cities began to grow around the factories and industrial centers. The Industrial Revolution and changes in the way people lived occurred rapidly.

At the beginning of the Industrial Revolution, coal and wood were the major sources of energy. Steam engines required large amounts of fuel to provide the heat for the boilers. In England many forests were cut down completely to provide the fuel.

In 1859 the first oil well was drilled in Titusville, Pennsylvania. It was soon discovered that the crude oil pumped from such wells could be made into fuel to fire the boilers of steam engines. Smaller engines could be made. The new engines were lighter and could be moved easier. The smaller, light-weight engines led to the development of better transportation. Trains, automobiles, and ships could use these engines. The first powered, heavier-than-air airplane flown by the Wright Brothers was powered by a gasoline-fueled engine.

As is evident in these examples of the past, technological advance depends on energy and power sources. Many of the problems we face today are related to energy and power.

Communication Systems

Great changes have also occurred in communication. Consider just three communication devices in your home—the telephone, radio, and television. When were these developed? The telephone was invented in 1875 by Alexander Graham Bell. The radio was invented in 1900, and the first commercial radio station went on the air in 1920. Television was not developed until the 1940s and 1950s. Ask your parents if they remember the first time they saw television. Ask your grandparents if they can remember their first radio.

Before the invention of these devices, communication was by word of mouth or by writing and printing. These were slow when compared to the telegraph, telephone, radio, or television. Although we still use writing and printing, we depend on faster means of communication. Faster means of communication have changed our lives. Today we can see events as they happen around the world. Our government leaders can talk immediately with leaders of other countries thousands of miles away. They do not have to wait days for a letter to be delivered and then wait days more for a reply. The computer has become an important part of modern communication.

This early telephone has slight resemblance to our modern telephone.

Transportation Systems

Our transportation systems are also dramatic examples of technological change. Until the invention of the steam engine, transportation was only as fast as a horse could run. It had been this way for thousands of years. The use of a steam engine to propel trains changed all this. The first steam train ran in 1825, slightly more than 150 years ago. Since that time transportation has changed so that today it is possible to travel thousands of miles per hour. Men have travelled into outer space and visited the moon. American astronauts rode across the surface of the moon on a *moon buggy* or lunar rover.

The invention of the steam engine changed the history of transportation. This train is emerging from a round house in Alexandria, Virginia in 1864.

Humans can travel on land, over it, and under the seas. Miners ride cars through tunnels miles underground to reach seams of coal. Modern submarines, powered by nuclear energy, can stay under water for months without surfacing.

Modern transportation technology makes it possible for strawberries grown in California to be eaten in New York the same day they are picked. Crude oil pumped from the ground in Saudi Arabia is refined in Pennsylvania. The gasoline produced in the refinery may be used as fuel in automobiles in the eastern part of the United States.

Production Systems

Changes in technology have had tremendous influence on production. Today we live in a mass-produced world. Nearly everything we use is mass produced. Sometimes hundreds, thousands, even millions of an item are produced. Each item is identical.

This was not always true. Before the development of interchangeable parts each item was different. Items were made one at a time. Skilled craftspersons designed, produced, and sold or marketed their own work.

Modern production systems use new materials, many of them products of technology. Complex processes are used to work these materials. In modern industries the flow of materials and

right The introduction of interchangeable parts on an assembly line was used by Henry Ford to produce motor cars. *above* Telephone assembly today.

Exploring Technology

parts is organized so that the items are made on the assembly line. Many of these production systems use computers to keep the system running smoothly. In some modern factories computer-controlled robots are used to produce items. Totally automated systems can turn out a product in only minutes without any worker touching it. The production line operates on an unmanned "lights out" basis!

Construction is a part of production systems. Houses are no longer built entirely by hand. Builders use many machines. Power saws make cutting lumber faster and easier. Tractors and back hoes can excavate for the foundation of a house in a few hours. It would take days using hand tools and horse-drawn, earth-moving equipment to excavate the same foundation.

Modern skyscrapers are possible because of new materials such as steel and synthetics. Modern machines such as elevators and cranes can lift the building materials. Concrete can be mixed on the ground and lifted to the top of the building under construction. The buildings are heated in winter by huge boilers and systems of pipes and radiators. They can be cooled in summer by piping chilled water through the system. Without modern construction technology, these buildings would not be possible.

How Does Technology Affect Us?

So far in this chapter you have seen that technology causes change. You have also seen some of these changes. It is obvious that these changes occur in many areas. In fact, they happen in all areas of human life. Technology affects people.

To better understand this effect on human beings the rest of this chapter will discuss how technology and people interact. It should help you develop a clear mental image of technology.

Changes in People

People are changed by technology. Medicines help us have better health. We are affected by fewer diseases. We are able to recover from illnesses more quickly. We are healthier because our foods are pure and safe to eat. The average person in the United States is taller and heavier than the settlers of the country. Medical technology helps people live longer. There are fewer deaths among children now because childhood diseases are controlled. Just a few years ago diseases spread through whole communities and countries, killing thousands. Elderly people also live in better health, and live longer. In America, the average expected life span is seventy-five or more years.

But, these technological advances are not worldwide. In some developing countries, the quality of life is much lower than in the Western world. Childhood diseases are not controlled. The people's diet is not as good. The food is not as pure nor is it as plentiful. Medical aid is not as good. In some places, it is not available at all.

Through technology people are changed in other ways. We have become accustomed to living with the products of technology. Our bodies are accustomed to life in warm houses. In the winter of 1976-77, many people in the United States turned down thermostats in their homes to sixty to sixty-five degrees to help conserve fuel. This was uncomfortable for some people who had grown accustomed to a temperature of seventy to seventy-five degrees in their homes.

We have come to depend on the automobile for transportation. People even ride short distances rather than walk. This product of technology, the automobile, has changed us until we depend on it. Our bodies are not accustomed to walking.

Technological change is not all good. Some of the products of technology can be harmful to humans. The automobile is a wonderful machine for transportation. It also creates air pollution, which can affect people's health. There are cases where air pollution was so severe as to cause people to die as a result of it. There are thousands of deaths each year as a result of automo-

Solar panels on the roof collect enough energy to heat the building.

bile accidents. These are some examples of the bad side of technology.

Insecticides, which are used to control harmful insects, also can have bad effects on humans. Farmers can grow larger crops and produce food that looks good by killing insects. The insecticides they use, if not used wisely, can cause illness in humans. The insecticides also can kill animals which eat the sprayed crops or insects.

Leisure Time

The work week gradually has become shorter. Tools and machines have made it possible to do more work in less time. People do not have to work as long each day. These changes are results of technological progress. Today a worker can produce more in a day than his grandparent could. This means that people have more leisure time. There is more time in each day to relax and enjoy the things around you. In 1900 a worker had to work sixty hours a week. Today it is forty hours or less for some.

Through technology many new leisure time activities have been created. Radio and television provide entertainment. Early radios were large and very heavy. Today a radio and even a television can be carried in a pocket.

People enjoy reading. Before the printing press was invented books had to be written by hand. There were few books and these were very expensive. Only the very rich could afford to read.

The bicycle is becoming an important transportation and recreation vehicle.

Look at the changes made in one of the popular leisure time activities—hiking and camping. Many people use modern camping trailers. The recreation vehicle industry is a multimillion dollar business. Backpackers use lightweight pack frames and nylon packs. Freeze-dried foods are light weight and taste nearly as good as fresh. A full dinner for four weighs about 1.3 kilograms (three pounds). People spend vacations traveling to visit distant parts of our country. Many travel to other countries. The automobile, airplane, train, and ship make these trips possible.

Our hobbies have been changed by technology. We have more time for hobbies, and we also have more hobbies. What is your hobby? Do you build models? Models are a product of technology. Airplanes, ships, trains, and cars are favorites of modelers. The plastic, metal, glue, and paint used in model building are products of technology. Are you a stamp collector? The stamps are printed on modern printing presses. These machines are very complicated and are marvels of technology.

The Environment

Technology is environment made by people. Technology is all around us. It is a part of our surroundings. Our houses contain their own environment. We heat them in the winter and cool them in the summer. Inside the house we create a mini-environment that is different from the out-of-doors. People can live comfortably in extreme climates such as Alaska or jungles or deserts.

Astronauts have traveled into outer space. This has been possible because they were able to take the earth's environment

An astronaut wears a spacesuit which provides an environment supplying oxygen and the necessary protection from the extreme temperatures on the moon.

with them. They carried oxygen, water, and food. Their spacecraft became a mini-environment.

Skylab, the space station or satellite, was like a mini-earth. The scientists were able to live and work in Skylab for weeks at a time. They also were able to go outside of the capsule and make repairs. They wore space suits that gave them the oxygen and protection they needed.

Technology affects us in another way. We are constantly exposed to the sounds, smells, and sights of technology. Some people complain about sound pollution. Some sounds are loud enough to damage our hearing. For many years factory workers suffered from hearing loss caused by noisy machines. Today many workers wear ear protectors.

In some areas of the cities signs and billboards line the streets and highways. Too many of these make the streets look cluttered and less attractive. This is called visual pollution.

We are all familiar with air pollution. Industries, power plants, and automobiles give off fumes that can affect our environment. "Smog alerts" are common in some large cities. The fumes collect until they are so concentrated that people become ill. "Acid rain" has become a problem in many areas.

Gaining Advantage over Nature
Through technology we can gain an advantage over nature. As mentioned earlier, people can live in arctic weather because comfortable houses can be heated. Farmers help control nature to suit the needs of humans by using many insecticides, fertilizers and hybrid seeds. These have been made possible by many technical advances.

The oceans were once a barrier to travel. Sailing ships and navigation methods overcame this natural limit to travel. Formerly, people could communicate only as far as their voices would carry. Smoke signals, the telegraph, radio, and television broke the distance barrier. Now we can see and hear events as they happen half way around the earth. We have been able to see the surface of our moon and other planets in our universe.

For thousands of years travel was slow. Speed was limited to that of a fast horse. The automobile, train, airplane, and rocket overcame the speed barrier.

Some attempts have been made to change the weather. There is some limited success at seeding clouds to make rain fall where and when it is needed. Fruit growers burn heaters in their orchards in an attempt to prevent the freezing of crops. Under

With the development of modern farm equipment, farmers can grow more food on less land with less human effort.

certain conditions water can be sprayed on crops such as strawberries to prevent freezing.

Changes in Work

As technology changes, work changes. There is more need for highly skilled workers and less need for the less skilled. Computer programmers were not needed a few years ago. Today many are needed. Just a few years ago people were needed to make horse-drawn wagons. Today there is little need for wagon makers.

Our complex technological society needs engineers, computer operators, and machine operators. Two hundred years ago there were large numbers of people working on farms. Today this number is much smaller. By using modern machines, seeds, fertilizers, and insecticides, farmers can grow more food, on less land, and with less human effort. Fewer people are needed to work on the farms. Much of the work is done by machines. Some tractors have air-conditioned cabs. A radio provides entertaining music or keeps the operator in contact with the home or other workers on the farm.

The changes in work are not all good. Work in many factories is monotonous. A person does the same job day after day, week after week. This can be boring. On the assembly line the worker sees only a small part of the work required to make an object. Hundreds of people do the work needed to assemble an automobile. Each person does only a small part of the total job. It is difficult for some of these workers to take any pride in their work. It is difficult to have a feeling of pride in the whole automobile when all one has done, day after day, is put on a wheel or weld on a fender.

Before modern mass production methods were developed, a worker often did the entire job. There was variety in the work. The workers could take pride in knowing that all the work was theirs. The final product was seen at every stage by the person making it. The assembly line-produced automobile is seen by only a small number of the people making it.

Another example of changes in work can be found in mining techniques. At one time coal miners dug the coal by hand using picks and shovels. In some early mines children carried or pushed the coal from the mines. Injuries and deaths of miners were common. Today modern coal mines use machines to dig the coal. Instruments are used to detect poisonous gases. The coal is transported from the mine in electric cars or on long conveyor systems. Mines are safer and result in fewer injuries and deaths. Modern technology is used to make the work of the coal miner less strenuous, safer and faster.

Materials

The human use of materials is an important part of technology. When you study the history of technology, you study the ways people have used tools to work with materials. The earliest tools of the stone age were made to chop and smash other materials so they could be used to make life easier.

Through technology we are provided with new materials also. Human beings were able to rise above mere survival by developing ways to work with the materials around them. New materials were developed from those found in nature. Copper, in pure form, is soft and easy to work using primitive tools. But, if melted and mixed with other materials, it becomes harder and more useful. Iron can be separated from iron ore and used in this form. But, when it is made into steel, it can be made harder or even rust resistant. Steels can be hardened so they will cut other metals, even other steels.

A sealer is applied to synthetic panels lining the floor of a huge containment reservoir.

Synthetics, or materials not found in nature, are products of modern technology. Scientists and technologists have discovered ways to make one material from another found in nature. Synthetic fibers and other plastics are made from petroleum. Some drugs and medicines also come from petroleum. Paper is made from trees.

Meat substitutes are made from soy beans. Synthetic strips of bacon taste similar to the real thing. Many foods are dried so they can be stored for long periods of time. Foods can be freeze dried for storage. When cooked properly, the taste is very nearly as good as fresh. The meals taken into space by the American astronauts were miracles of food technology.

Self-Expression

Through technology we are able to express ourselves in many ways. A sculptor can work with welded metals, synthetic clays, plaster, or molten metals. All are products of technology. The painter can use modern oils, water colors, or acrylics—also products of technology.

Optical art, artistry in lights, metals, plastics, and paper are possible because of technical advances. Movies, radio, and television are areas of possible expression by writers and actors. The politician can communicate ideas to millions of television viewers. The writer's ideas reach millions of people through printed books, magazines, and newspapers.

None of these is possible without the developments of technology. Early people were limited to rather simple ways of expressing themselves. Materials were limited to those found in nature. The sculptor was limited to wood, stone, or native clay. The writer was limited to writing for only a few rich people.

Cultural Changes

Changes occur in the social, economic, moral, and political aspects of human life. In some countries most people can read, no matter how poor they are. Modern printing methods make books cheap enough for all to afford. As people learn to read, they can understand better the world around them. One of the first things missionaries often do when going to a primitive country is to teach the people to read. The people are then able to learn about improved ways of living—better medical care, better foods, and how to prepare them. They learn to take better care of their families. The radio can reach many people who cannot read. There are still millions of people on earth who lead a rather primitive life. Their living conditions can be improved through the use of technology.

The United States, Sweden, Denmark, Japan, and West Germany are examples of countries with high standards of living. Millions of people living in less developed nations do not enjoy this life. In about 400 years our ancestors have raised our level of technology to unbelievable levels. The developing countries will not have to wait that long. They can learn from us and adopt the higher levels of living. They can see the higher level of technology and begin using it. They can raise their level of technical development in less than 400 years.

Model of Technology

Though not mentioned before in chapter one, the explanation of technology given here is based upon a concept of what technology is. A diagram or *model* of technology can be drawn to show this concept.

Perhaps the use of the word model should be explained. Have you ever built a model of an airplane, ship or a car? What was this model? It was a miniature of the real thing. When people look at your model they can get an idea what the real airplane, ship or car looks like, even if they have never seen one.

People looking at the model on page 28 can get an idea what technology is. Just like the airplane model, the model of technology is made up of many parts. Each part represents a part of the

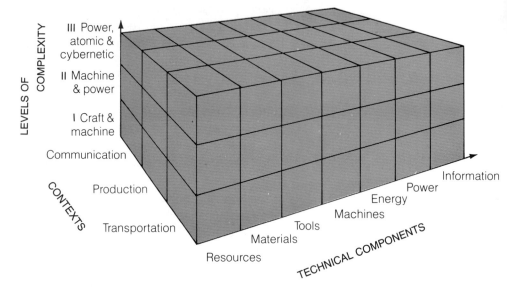

real thing. When all are put together, the model represents all of technology.

This model of technology helps to explain several of the important parts of technology. You will notice that most of these have been discussed in this chapter.

First, let us look at the *contexts* of technology. These are listed on one of the sides of the model. Humans have developed technology in three *contexts* or areas. These are production, communication and transportation.

Communication has always been an important human concern. People have developed many ways to communicate. This is one of the things that makes people different from other creatures on earth. We have developed ways to communicate other than just talking. These are our technology of communication.

Production includes processing, manufacturing and construction. Humans are producers. Raw materials are taken from the earth and processed into a usable form. These materials are then used to make products such as cars, toasters and bicycles. Today this is usually done in a factory using mass production techniques. Some materials are used to build bridges, houses and roads. We usually refer to these as products of construction.

From primitive sleds to modern spacecraft, people have tried to improve their means of transportation. Faster ways to move people and materials have been developed. A technology of transportation was developed.

While developing better ways to communicate, produce and transport, people developed other things. *Resources* such as oil, coal and iron ore had to be discovered and developed. These had to be processed into *materials*. *Tools* and *machines* had to be developed to change the materials into usable products. *Energy* and *power* sources had to be developed to make all this easier for people to do. In communication, *information* systems were developed. Systems of writing and calculating were developed. Today information is processed by computers.

The above are called the *technical components* of technology. The components are parts of technology that make it possible for people to communicate, produce and transport.

There are also social and cultural components or parts of technology. People are affected by technology. Work, leisure, and our relations with other people are affected. These are shown as the social components or technology.

The components of technology are listed along another side of the model.

As you read this book you will learn about the historic development of technology. You will learn that as technology developed it became more complicated or more complex. The technology of the early stone age people was very simple. It was much simpler than that of the settlers of this country in the seventeenth century. Today our technology is much more complex than ever before.

We can say there are three levels of complexity of technology. These are shown on another side of the model. These are (I) craft and machine, (II) machine and power and (III) power, cybernetic and atomic.

Level I is the simplest or least complex. It includes the age of human development before the age of complex machines. During this time work was done primarily by hand. Humans and animals were the main sources of energy.

Level II is the time of powered machine development. The Industrial Revolution is a feature that stands out in this level. People learned to add power to their machines. The machines and people were organized into factories. Mass production was born and grew during this time.

Level III is our modern level of complexity. We live in an age of atomic power, of automation and of cybernetics. Computers are used to extend our brains. Machines have made life easier for humans.

Would you like to predict the future level of complexity? How complicated could our lives become in the future? Is there a limit to this increasing complexity? Sometimes it is fun to think about the future. It can be fun to try to guess what our future will be like. Maybe if we think about our future we can be better prepared for it.

There are several ways you can study technology. One is to study one of the contexts at a time—transportation, for example. You can narrow down your study even more. You can study just one of the levels of complexity. The first level, craft and machine, is an interesting level for study. You would then be studying only transportation as it was before the use of engines and other sources of power.

You could further narrow your study by focusing on only one of the technical components of technology. Machines are a good example. You would then be studying the machines used in transportation during this early time in history.

Can you follow this on the model? If you narrowed down your study as described above, which cube would represent your area of study?

Home computers are now common. They can be used by all family members for homework, games, accounting and writing. Computers enliven many of these tasks.

Suggestions for Further Study

OBSERVATIONS

There are many ways you can become involved with technology. Museums can be a good source of information. Most museums have displays that relate to some area of technology. Some of the museums are designed just to display technology of the past. There may be one near your home that will help you in your study of technology.

Look around your community. Is there some display or museum currently on exhibit? Many local museums are interesting to visit. They help you understand the history of the area in which you live. Many industries have displays in their office buildings or other areas open to the public.

You can learn much about technology through hobbies. Building models of airplanes, ships, trains, cars, and rockets is a favorite activity. Throughout this book model making will be suggested as a good way to learn more about technology. When you make a model of an object you are learning about that object. You learn how ships are built when you build a ship model. Model airplanes are fun, especially when they fly. You can learn much about how an airplane flies when you construct a flying model airplane.

Constructing models of buildings is a good way to learn how a house, for example, is put together. Other structures such as bridges can be entertaining to build as models.

Questions

1. What is technology?
2. How does technology affect mankind?
3. Can humans control nature? How?
4. How has technology changed communication, transportation, and materials?
5. How has technology changed the way you use leisure time?
6. What changes do you predict will occur in the future as a result of technology?

Key Words

technology
synthetics
technological change
tools

machines
communication
transportation
production

Developing a Time Line

Concept: *Change Over Time*

Many technological changes have occurred over the thousands of years of human existence. Studying the history of technology is important to understanding the how, where, and when of technological developments. This understanding helps us to appreciate the importance of new developments. It also helps us to predict future developments.

Facts are like building blocks. Each builds upon those below and supports the ones above. A technological development is not a single event. Many earlier events made this new one possible. This new event will make newer events possible in the future.

Objective

In this activity you will gain a better understanding of the history of technology by developing a time line. You will learn some of the ways technological developments are related to each other and to other historical events.

Procedure

To begin your time line, draw a line through the center of a piece of notebook paper. Divide this line into ten parts. Label each part with dates of the time period you want to study. The time line in the illustration has been labeled for 1800–1850, covering a period of fifty years.

Label the columns to the right and left of the line for the technological systems to be studied. The example has been labeled "energy/power" and "transportation."

As you discover important technological developments in each of these systems, list them at the appropriate place along the time line. Note that in the example entries are listed for 1800, 1821, 1826 and others in the "energy/power" column.

When complete, your time line will let you compare developments in the two systems. You can also compare developments in other fields. To do this add a third column labeled "other events." In this column list events not considered to be technological developments. For example, in 1812 the U.S. Congress declared war on Great Britain. Did you know that important developments in transportation were taking place at the same time as the War of 1812?

Have each person in your class study a different time period. (1700–1750, 1750–1800, 1800–1850, for example). When completed, the time lines can be connected and displayed on the wall of the classroom.

Make the time lines more interesting by adding illustrations or original art work. Copy or xerox illustrations from books and magazines and paste them onto the time lines. (DO NOT cut illustrations from the books.)

Technological Progress
1800–1850

Energy/Power		Transportation
	1800	
1800 Electric battery invented by Allessandro Volta		1807 Henry Bell (Scotland) designs first European steamship
		1807 Robert Fulton's Steamboat "Clermont" travels up Hudson River
		1809 First ocean-going steamboat, "Phoenix"
	1810	
		1818 Transatlantic ships begin regularly scheduled trips, Liverpool to New York average time of trip — 30 days
		1819 Macadam roadmaking technique introduced by John McAdam (Scotland)
	1820	
1821 Electromagnetic rotation discovered by Michael Faraday		1825 Erie Canal completed
1826 Internal combustion engine patented by Samuel Morey		
1828 Electromagnet invented by Joseph Henry		
	1830	
1830 Electromagnet induction discovered by Henry — electricity generated		1832 Horse-drawn streetcars used in New York City
1831 First electric motor built by Henry		1835 America's first cast iron bridge built
1837 Electric motor patented by Thomas Davenport		1836 Screw propeller with blades invented by John Ericson (Sweden)
		1839 Vulcanized rubber produced by Charles Goodyear
	1840	
	1850	

Ore deposits such as coal that lie near the surface are extracted using open pit mining techniques. The smaller machine in the foreground breaks up the deposits. The hugh crane in the background runs a dragline bucket for collecting ore and debris.

Chapter 2   Processing Technology

Introduction

HISTORICAL DEVELOPMENT OF PROCESSING

Time	Development
1635	Chemical plant opened in Boston — first in U.S.
1715	Iron Plant established near Fredericksburg, VA
1750	First American coal mine opens — Virginia
1784	Puddling process of iron production patented in England
1839	Vulcanized rubber produced by Charles Goodyear (patented in 1844)
1855	America's first oil refinery — Pittsburg, PA
1857	Bessemer (England) develops steel making process using blasts of air to remove impurities
1859	Edwin Drake drilled America's first successful oil well

continued

What is processing? Possibly the best way to answer this question is to look at several examples of processing. Through these a definition can be learned.

In early history, people used materials as they were found in nature. These early people did not know how to change the natural materials in ways that would make them more useful or easier to use. The Stone Age people learned to use stones as weapons when killing animals for food. Later the stones were tied to the ends of sticks. This made a club that would create a harder blow. Though this simple change was important in the early history of technology, it was not processing technology. A tool had been made but there was no change in the raw materials being used. Two objects, the stone and the stick, were merely combined. These early people did not have a processing technology. It would be thousands of years before the technology of processing would be developed.

On some parts of the earth, crude oil could be found oozing out of the earth. The people who lived in these parts of the world found they could use this material to make things slide more easily. They only knew how to use the oil as they found it coming out of the earth. They did not know how to change it into something more useful. They didn't have a processing technology.

Early people used plants and animals for food. This food was often hard to find because these people could only use what they found in nature. What food they found was eaten raw as they found it. They did not have a processing technology. When these

HISTORICAL DEVELOPMENT OF PROCESSING

Time	Development
1861	Open-hearth furnace heated by gas is patented
1896	Successful off-shore oil wells drilled near Santa Barbara, CA
1899	Steel is heat treated to increase its strength and cutting ability by 300%
1909	Duralumin—strong, lightweight aluminum alloy developed—later radically influences evolution of aircraft
1913	Process for converting coal dust into gasoline developed in Germany
1928	First electric iron-smelting furnace
1934	Nylon produced
1935	U-235, an isotope of uranium is discovered—later used in nuclear power plants and weapons
1936	Catalytic cracking introduced for refining petroleum
1943	Polyethylene plastic introduced
1954	Basic oxygen process of steelmaking introduced in U.S.
1968	Spray steel-making developed in England
1986	Voyager—all composite aircraft completes around-the-world, non-stop flight without refueling

early people discovered that this food could be cooked before being eaten, they began to develop a processing technology. They were learning to change the form of the food to make it taste better or to make it more edible in some way.

Over many thousands of years these early people learned that many materials they found in nature could be changed in some way to make them more useful. Pieces of flint could be shaped in a special way and tied to a stick to make an arrow. The piece of flint was being changed from its natural shape into a shape that was more useful. This was processing technology.

It was discovered that the hides of animals could be scraped, treated in a special way, and dried. This made the hide last a long time and made it useful for many things. The tanning of this leather is another example of processing technology. The hide as found in nature was changed into a more useful form.

Early people, then, slowly learned to change the form of materials they found in nature to make these materials more useful. We call this processing. By modern standards these early examples do not seem very dramatic, but they are examples of the early start in the technology of processing.

Why is this important? It is important because today few materials are taken from the earth and used directly as they are found. They are processed in some way to make them more useful. For example, crude oil can be refined into gasoline, fuel oil, kerosene, and hundreds of other useful products. Cotton and wool fibers can be made into thread so that cloth can be woven from it. Trees are pretty to see, but they can also be cut and made into lumber or paper. Houses, barns, churches, and bridges can be built from the lumber. Books can be printed on the paper.

In these modern examples, materials found in nature are being changed in some way to make them more useful. This is processing. Processing, then, is the changing of raw materials to make them more useful.

From the examples given above, you can see that processing is very important. Without modern processing techniques, many of the materials we use and need would not be available. There are many instances in the history of technological development where progress was delayed until a necessary new material became available. Many of the tools we use were not possible until steels to make tools were developed. Modern space travel was not possible until materials were developed for use on the heat shields of the reentry vehicle. Without these materials the reentry vehicle would burn up as it entered the earth's atmosphere.

Cotton yarn is wound onto spools and then twisted into the threads used to weave cotton fabric.

To make the study of processing easier, let's organize it into several smaller parts or ideas. Each smaller part can be studied, then the parts can be put together to help understand the bigger idea, processing. This is like studying a jigsaw puzzle where one can lay out the parts, then put the parts together to see the total picture.

Processing technology can be studied by looking at three smaller parts. These are preprocessing, processing, and postprocessing. This table illustrates these three parts.

TYPICAL OPERATIONS IN PROCESSING TECHNOLOGY

	Typical Operations
PREPROCESSING (Before Processing)	Locating Harvesting Drilling Extracting Storing Transporting
PROCESSING	Refining Reducing in size Separating Conditioning Forming Combining
POSTPROCESSING (After Processing)	Storing Transporting Distributing

Processing Technology

Preprocessing refers to all the operations that must be done to a material before it is processed. (Pre-means *before*. Preprocessing, then, means *before* processing.) This includes all the operations that are done to a material before it is changed into useful form, or processed. These include finding the material, taking it from the earth as in mining and harvesting, and moving or transporting it to the processing plant.

Processing takes place in a plant of some kind. Processing plants include oil refineries, leather tanneries, lumber mills, paper making plants, food processing plants, and many others.

Postprocessing refers to what is done to the material after it has been processed. (*Post*-means *after*. *Post*processing, then, means *after* processing.) Processed materials must be moved from the processing plant to the places where they are to be used. In many cases the materials must be stored until they are needed. Sometimes the materials must have extra processing before they can be used.

Preprocessing

Preprocessing can also be organized into several different areas. Four are used here: sources of materials, extraction of materials, storage of materials, and transportation of materials.

Sources of Materials

The earth is the source of all our raw materials. Although the astronauts brought back rocks from the moon, the moon is not a practical source of raw materials for use here on earth. Can you imagine what a bicycle would cost if the iron used to make the frame had to come from the moon!

Raw materials are found all over the earth. Some are found in larger amounts in some places than in others. These larger concentrations of materials make it possible to remove them with minimum cost. Aluminum is a good example. Aluminum is made from bauxite ore. Some bauxite ore can be found in practically all parts of the earth. But, it is economical to mine the ore only in those places where it can be found in concentrated form. These places are where the bauxite mines are located.

These areas of concentrated ore must be located before the ore can be mined. But, just locating the material is not enough. It also must be easy to remove. It must be readily available. For example, crude oil can be found under the ocean in many places. But it can be removed easily from the areas where the depth of the

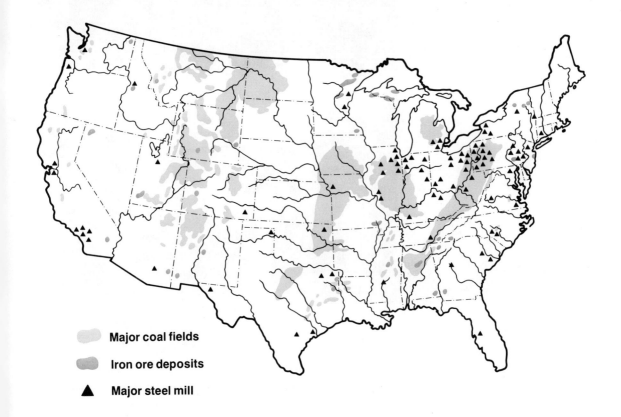

- Major coal fields
- Iron ore deposits
- ▲ Major steel mill

ocean floor is not too great. To drill for oil in the deep parts of the ocean would cost too much.

Oil was discovered on the north slope of Alaska. This oil was not readily available until a pipeline was built to transport it to a shipping port. The oil was there but it was not readily available.

Once found, the material must be transported to the processing plant. Often processing plants are located near the source of the materials to reduce transportation costs. The plants may be located near some available source of transportation. This is the reason many steel mills are located near rivers or lakes. The coal, limestone, and iron ore needed to make the steel can be brought to the mills in ships.

Many lumber and paper mills are located in the northwestern and southern parts of the United States. Large forests are located there. The processed lumber or paper can be shipped wherever it is needed. It is cheaper to transport the processed materials than to transport the logs.

Recycling is one way to conserve our limited resources. These compacted steel cans will be melted down and the steel reused.

The supply of some materials is limited. These are called *nonrenewable* resources. There is just so much of these materials on earth. When they are used, there is no way to replace them. Coal, oil, metal ores, and natural gas are examples of nonrenewable resources. These materials were formed in the earth over a period of millions of years. Though there seems to be plenty of oil now, remember that each day there is less than the day before. Some of this nonrenewable resource is being used each day. Is it being used in the most efficient way? Is it being used for the most important reasons? Is some of it being wasted?

Some natural materials are *renewable*. This means they can be replaced as they are used. Trees, for example, are renewable resources. As they are cut, more can be planted to replace them. Animals such as cattle, sheep, and hogs are renewable resources. Animals can be grown to replace those that are slaughtered for food.

Careful use must be made of even our renewable resources. Some take a long time to grow. A tree can be cut in minutes. It may take thirty or forty years for the replacement tree to grow. The redwood trees took hundreds of years to grow. It would take an equal length of time for new ones to grow to replace those cut for lumber. If managed properly, we should not run out of the renewable resources.

Recycling is one way to conserve our limited resources. By recycling, using materials several times, less new material need be found. This helps slow the use of new materials. There is much interest in recycling aluminum. Aluminum companies collect aluminum cans to be melted and used again. In addition to saving some of the limited supply of bauxite ore, recycling saves a large amount of energy. Recycling aluminum uses only 5 percent of the energy required to get an equal amount of aluminum from the ore.

Trash that is collected in the cities can be recycled. Each year billions of tons of this trash are collected and burned or buried. It contains paper, cans, glass, plastic, and many other usable materials. It is estimated that about one half of the trash from a city is paper, one fourth is metal, and one tenth is glass. Much of this can be recycled and used again. Recycling these materials would save many of our natural resources.

Recycling paper would reduce the number of trees being cut. Recycling plastics would save petroleum. Scarce ores could be saved by recycling metals.

Extraction

Once the raw materials are found they must be removed, or extracted, from the earth. Mining is one of the methods used for extracting materials. Liquids such as oil and gas are extracted by drilling into the earth. Trees and other plants are harvested.

Mining is the process of digging into the earth to remove rocks and minerals. The mining method used depends on the material being mined and its location in the earth. Some materials, such as coal, are found in a compact mass. Others, such as bauxite ore are scattered.

Surface Mining. When ore deposits lie near the surface of the earth, surface methods are used. These methods include placer mining, dredging, open pit, strip mining, and quarrying.

Placer mining was the method used by many of the miners of the gold rush days in the United States. The gold was found mixed in gravel deposits along stream beds. The gold-bearing gravel could be washed with water to separate the gold from the other materials. Gold is heavy enough to settle to the bottom of the container, allowing the other materials to be washed away. Using a shallow pan, the prospector scooped up a small amount of gravel and water. He swished this around in the pan. The gravel and dirt was washed over the side leaving the heavier gold in the bottom.

Larger placer mining operations used riffle boxes to remove the gold from the gravel. These were long wooden troughs with strips of wood fastened to the bottoms. Water ran through the troughs and ore-bearing gravel was shoveled into them. The water washed away the gravel, leaving the heavier gold trapped in the strips in the bottom of the trough.

Hydraulicking is a method of placer mining where high pressure water hoses wash the gravel directly from the ore deposit into the riffle boxes.

Open pit mining is used when mining minerals near the surface of the earth. Often whole mountains are removed in order to reach the minerals. Ledges or terraces are cut around the sides of the pit. These form a roadway for trucks and earth moving equipment. Holes are drilled into the rock in the sides of the terraces. Blasting material is placed in the holes and the sides of the terraces blasted down onto the terraces. Power shovels then are used to load the ore into trucks or trains for transportation to the processing plant. Some of the larger trucks can haul as much as 100 tons of ore.

Nearly all (90 percent) of iron ore is mined in open pits. Natural deposits of iron ore are being depleted so there is increasing emphasis on the use of lower grade ores. This means it is harder to mine the ore inexpensively and easily.

Iron ore from this open pit mine in Minnesota's Mesabi Range is loaded into railroad cars for transport to the surface.

In *strip mining* the earth covering the mineral deposit is removed before the ore is mined. When the deposit lies near the surface, earth is removed to expose the vein of mineral. The removed earth is called overburden.

There are three kinds of strip mining: area, contour, and augering. Area mining is usually used on flat terrain where giant drag lines and shovels strip away the overburden. The exposed mineral can then be removed.

Contour mining is used in mountainous areas. Cuts are made into the mountain side to uncover the mineral. The ore is removed then by using large power shovels and other earth-moving equipment.

When mining by the augering method, a shelf is cut into a hillside. Large bits or augers are placed on the shelves. Holes are drilled into the seam of mineral, coal for example, taking out as much as possible.

When there are layers of mineral-bearing sand or minerals under water, they are removed by *dredging*. The dredge is mounted on a floating barge. Buckets in an endless chain are lowered to the bottom of the lake or river where they dig up the sand or gravel. This is washed and the unwanted materials are returned to the water. The desired materials are transported to the processing plant.

Draglines or slacklines operate similarly to dredging. Instead of an endless chain of buckets, one large bucket is pulled through the deposit. The mineral is lifted up and dumped on the deck of the barge. After washing, the mineral is transported to a processing plant.

In *quarrying*, large blocks of rock or mineral are broken loose, usually by blasting. Some materials are first sawed into blocks, then broken loose. Marble, granite, sandstone, limestone, and slate are some of the materials mined by quarrying. The quarry is similar to an open pit mine.

Underground Mining. Minerals that lie deep within the earth must be mined by underground methods. These include slope mining, shaft mines, and drift mines. *Slope* mines are used frequently to mine coal in hilly areas. A sloping tunnel is dug to reach the coal bed in the hill. Slope mines are used when the coal lies near the surface but too deep to be reached by surface methods. The miners can ride or walk down the sloping tunnel to reach the area where the coal is being dug. Cars or conveyors carry the coal out of the mine.

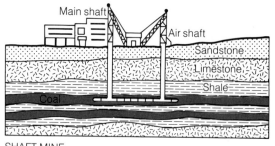
SHAFT MINE

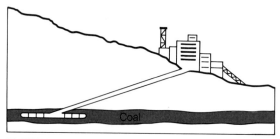
SLOPE MINE

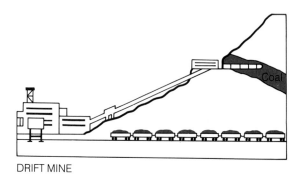
DRIFT MINE

Underground Mining Methods

Shaft mines are used to mine several different minerals. Copper, lead, zinc, gold, and silver as well as coal often lie deep below the surface of the earth. These are reached by digging a shaft straight downward to the vein of ore. From the vertical shaft, horizontal tunnels are dug into the vein. Conveyors carry the ore to the vertical shaft where it is lifted to the surface in elevators. These elevators are similar to the ones in high-rise buildings used to transport people.

In shaft mines it is often necessary to dig several shafts. Some are used for ventilation to get fresh air to the miners working underground. Large fans are used to force air down the shafts into the tunnels and to the work area. The movement of air also helps take out any poisonous gases that might collect.

Shaft mines are often as deep as 75-85 meters (250-300 feet). Some are as deep as 450 meters (1,500 feet). In Europe some shaft mines extend to a depth of 1,200 meters (4,000 feet).

Drift mines are used to remove coal that is found in hillsides. Often the seam of coal can be seen on the surface of the hill. The

Rotary Drilling Rig

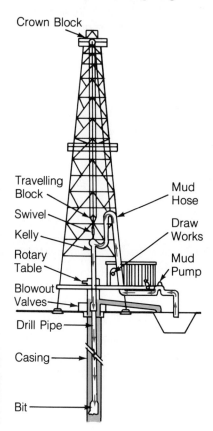

entrance to the mine can be made by simply digging into the exposed vein of coal.

Electric-powered cars haul miners into the mine. Conveyor belts are sometimes used to haul the coal. The drift mine is more economical to dig than the shaft or slope mines. Shorter tunnels and fewer expensive elevators and conveyors are needed.

Drilling. Oil, water, and gases are removed from the earth by drilling, one of the extraction processes. In drilling, a hole is made into the earth and a pipe is inserted to remove the liquid or gas.

Rotary drilling is the most important method in use today. A bit is attached to the lower end of a string of drill pipes. The bit and pipe are passed down through a rotary turntable in the floor of a derrick. The pipe and bit are turned as they are lowered into the earth. The bit cuts deeper and deeper into the earth. Pieces of drill pipe are added to the top of the string as it lowers into the earth.

When the bit breaks or becomes dull, the workers must pull all the pipe from the hole, section by section. As each section is lifted, it is removed from the string and laid aside. Sometimes several thousand meters of pipe must be removed from the hole.

After changing the bit, the workers reassemble the string of drill pipe, section by section. Drilling can begin when the bit reaches the bottom of the hole.

Drilling requires highly skilled workers. The *driller*, who heads the crew, directs the work. The workers are called *roughnecks*.

Oil workers guide a pipe through a rotary table.

Processing Technology

Drilling goes on day and night. Crews usually work in eight-hour shifts. Skill and extreme care must be used to prevent breaking the string of drill pipe. If a break occurs, there is great loss of drilling time. If a broken piece of pipe cannot be removed, the well will be lost. This could require moving the rig and starting a new well.

Cable tool drilling is simpler than rotary drilling. A hole is punched into the earth much as a person would punch a hole by driving a stick in with a hammer. A bit is attached to the end of a metal rod. (This is called a stem.) This, in turn, is attached to a long cable. By pulling up the cable, the stem is raised. It is then dropped to punch into the earth. This raising and dropping of the bit is done time after time. Each time it is dropped, the bit punches deeper into the earth. The bit has jagged teeth, which crush the soil and rock beneath it.

Even in cable drilling, the bit must be removed from time to time so the hole can be cleaned. Often a new sharp bit is put on. Cleaning is done by forcing water down into the hole. A long pipe called a bailer is used to pump out the water and bits of rock and dirt. Once the hole is cleaned, the drilling tool is put back into the hole and drilling continues. Drilling and bailing continue until the well has been drilled to the desired depth.

The first oil wells were drilled using the cable-tool method. It is seldom used today for oil drilling. Shallow water wells are still drilled using the cable-tool method.

Storage
Once the raw material has been extracted, it is often stored until it can be moved to the processing plant. Crude oil and gas are stored in tanks near the oil or gas fields. Minerals, including coal, iron ore, and limestone, are simply stored in piles. Trees that have been harvested also are stored in piles until shipped. Sometimes logs are stored in ponds or lakes.

Some materials must be protected from the weather during storage. In such cases specially designed buildings are used for storage. The grain elevator is an example. Grain such as corn and wheat must be protected from the weather to eliminate spoilage. Foods are stored in refrigerated buildings. Some fruits are stored in buildings in which a special atmosphere is created.

Transportation
The raw materials must be transported to the processing plant. Liquids are carried in pipelines, tank trucks, train cars, or ships.

A milk truck arrives at a dairy with 5600 gallons of milk to be processed.

Some of these tankers must be specially designed to carry special materials. A perishable liquid such as milk must be refrigerated to prevent spoilage. The milk tank is usually made of stainless steel to prevent contamination of the milk. Other materials must be heated so they will not harden. Asphalt is an example.

Solid materials such as iron ore and coal can be carried in open-hopper train cars or trucks. Special ore boats are built to carry the ore and similar solid materials on water.

Coal, though a solid, can be transported through a pipeline. If finely ground and mixed with water, the coal slurry can be pumped through the pipe. When it arrives at the end of the pipe, the water is drained and the coal is dried before using.

Trucks are used in modern logging operations to carry the logs from the forests. If the lumber mill is located near a body of water, the logs can be transported on ships or barges. Trains can be used to carry logs to the mills in certain areas.

Some materials are stored after reaching the processing plant. They are moved directly from the place they are extracted or harvested to the processing plant. Some of these must be stored at the plant until ready for processing. It is important that there be a steady flow of materials into the plant. Any shortage of materials will cause costly delay in the processing. Delays, for any reason, can increase the cost of processing the material. For this reason, many processing plants stockpile some raw materials. When there is a delay in transporting materials to the plant, the stockpile can be used. When the material is flowing once again, the

Processing Technology

stockpile can be replenished. In this way there is a steady flow of material into the plant. Processing is not interrupted without a costly loss of time.

Processing

Many methods exist to change raw materials into a more usable form. Here you will study several of these to process steel, paper, lumber, and other materials. The table below shows basic processing operations. As you read about processing in the rest of this chapter, see if you can find where these are being used.

Many raw materials are *reduced in size* during processing. They must be made smaller before they can be used. Most ores are crushed. The large pieces from the mines are too large for the processing machines. Coal comes from the mine in large chunks. These are crushed into different sizes as needed for various uses.

Other materials are cut into smaller sizes. Trees are cut into lumber. Pulpwood logs are chopped into chips. We don't usually

BASIC PROCESSING OPERATIONS

Operations	Examples
Reducing in size	Crushing Grinding Cutting
Combining/Mixing	Mixing Blending
Separating	Filtering Sifting Refining Drying Distilling
Grouping & Classifying	Size Color Kind Shape Quality
Conditioning	Hardening Softening Heating Cooling
Forming	to shape to size

think of it in this way, but food is reduced in size before it is processed. Beans and strawberries are cut into smaller pieces before being frozen. Meat is cut into steaks and roasts. Corn is cut from the cob.

Some materials are ground into a very small size. Lime and cement are made by grinding limestone rock into a fine powder. Coal is ground into a powder so it can be blown into furnaces and boilers.

Raw materials also are *combined* during processing. By putting together or combining two or more materials, a new material can be produced. Solder is a common example. Lead and tin are combined in certain amounts. This mixture, called an *alloy*, has special properties not found in either of the materials alone.

Pewter is another example. Early metalworkers used pewter in fine work such as pitchers, goblets, and plates. Modern pewter is an alloy of tin, antimony, and copper. Combining the three metals in certain amounts produced the new material, pewter.

It is often necessary to *separate* materials. In coal mining, rock must be removed from the coal. In paper processing, the bark must be removed from the pulpwood log. Oil refining is a process of separating the various materials from the crude oil. Solid materials must be filtered from many liquids. Cream is separated from the raw milk before butter can be made from the cream.

Some raw materials must be *grouped or classified* before processing can be done. Grouping may be done by color, size, or quality. Logs are grouped by hardwoods or softwoods. In a lumber mill, the walnut boards are grouped separate from the maple. Small logs may be used in paper making and the larger ones cut into lumber.

Conditioning is required before some materials can be used. A well-known example is the hardening of steel (tempering) by heating and cooling it in a certain way before it is made into a cutting tool. Lumber is dried before it is made into furniture. Many foods are cooked before we eat them.

Some materials are *formed* into shapes so they can be used later in manufacturing. Some iron ore, for example, is formed into pellets to be shipped to steel mills.

In the rest of this chapter several examples of processing will be discussed in detail. This should help you understand better the many processing methods. Notice that several methods are used in converting each raw material into usable form. Rarely can only one method completely change a material from the raw state into a form that can be used.

Raw pineapples are *reduced* before they are packed into cans.

Processing Technology

This is just one part of a huge oil refinery where crude oil is processed into many useful products.

Processing Petroleum

Refining. Crude oil in its natural form is not a good fuel. It is a sticky, gummy mass that has little use until it is processed. Many useful materials can be separated from the crude oil. This processing of crude oil is called *refining*.

At first glance a refinery is a maze of tanks of many sizes and shapes, connected by endless miles of pipes. Someone said a refinery looked like a modern sculpture of spaghetti and meat balls. Actually a refinery is a marvel of sophisticated design. In most refineries, computers are used to help operators keep track of what is going on inside this tangle of pipes and tanks. Dials on control panels tell operators the temperatures, flow rates, and pressures at hundreds of points in the pipes and tanks. Remotely controlled valves keep the crude oil moving smoothly into the refinery and products moving from it in a continuous stream. Samples are taken at many points to help make sure a high quality product is produced.

Fractional Distillation Tower of an Oil Refinery

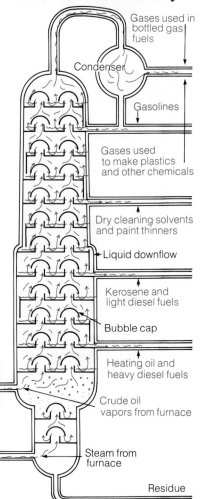

Crude oil goes through a variety of processes as it is refined. Some of the processes change heavier parts of the crude oil into lighter, high quality gasoline, aviation fuels, heating oils, and diesel fuels. Heavier materials from the refining processes are made into products such as asphalt.

Not all refineries turn out the same products. Some produce solvents, waxes, greases, or lubricating oils; whereas others produce the raw materials needed to make plastics. Each refinery is custom made to handle certain types of crude oils and to produce certain products. Crude oils that come from different parts of the world are different. For example, the crude oils from some parts of earth result in gasoline and fuel oil more easily than crude oil from other places. Nature puts only a certain number of gasoline producing molecules in crude oil. Thus some crude oils may yield more gasoline than others when refined. Other crude oils may yield more waxes.

Different crude oils also contain different amounts of natural elements that are not wanted. Sulphur is an example. A crude oil that contains little sulphur is called "sweet" crude. A crude that is high in sulphur content is called "sour." This sulphur causes problems for the refiners. The sulphur corrodes metal and must be removed to keep it from polluting the environment. Special metals must be used for constructing the tanks in refineries that process "sour" crude. Special machinery must be built to remove the sulphur. This means that a refinery is usually designed to process either sweet or sour crude, but not both. Some refineries are built to run both types of crude through different units.

In modern refineries the crude oil is pumped through a series of pipes, tanks, and furnaces where it is heated, often to as high as 250° C (725° F). This *conditioning* creates a hot mixture of liquid and vapors. These are pumped into the bottom of a tall tank called a fractionating tower. This tower is often as much as 30 meters (one hundred feet) tall.

As the vapor from the heated oil rises in the tower, it condenses, or turns to liquid, on plates called "bubble caps." The liquid that forms on these plates is different at different heights. It is collected and drawn off for various uses. This is a *separating* process called distillation, in that some materials are separated from the heated oil. High up the tower kerosene and light diesel oils form. At a higher level paint thinners and dry cleaning solvents are collected. Near the top of the tower the lighter materials used to make gasoline form and are separated.

Processing Technology

Some vapors do not condense. These are piped out of the top of the tower and used to make propane and butane fuels. Gases used to make plastics and many chemicals are also collected at the top of the tower.

Some materials stay in the bottom of the fractionating tower. These are distilled again in a low pressure process to produce lubricating oils. Some materials still remain and are treated to make asphalt, which is used to make floor tile, roofing materials, and materials used in paving roads.

Distilling processes other than fractional distillation are used. Thermal cracking and catalytic cracking break down some of the heavier materials in the crude oil that do not break down in the distillation process. This increases the amount of gasoline that can be obtained from a barrel of crude oil. It also produces a higher quality gasoline.

Petroleum Products. Gasoline is the chief product of petroleum. Automobiles, buses, trucks, and tractors use most of the gasoline in the United States. Fuel oils rank second to gasoline in importance. They provide a major source of heating and power. Fuel oil is used to heat homes, schools, factories, and other buildings and to supply power to factories, railroads, ships, power generating plants, smelters, and so forth. Jet fuels can be a special mixture of gasoline, kerosene, and oils with low freezing points. Straight kerosene can also be used as a fuel.

Lubricants or oils and greases are used to provide a slippery film between moving parts of machines of all types. Without this film the machines would quickly stop from overheating and friction. Close fitting parts would stick together without this slippery film. Although they make up a small portion of the total use of petroleum products, lubricants could quite easily be the most important use. Without lubricants, machines or factories, automobiles, even the machines that generate electrical power would not run. Without lubricants, our whole technological society would come to a stand still.

Asphalt and road oil are used to surface roads. Large amounts of asphalt are used to cover roofs of buildings, particularly factories and other large commercial buildings. Shingles used on homes are made from asphalt. In the United States 90 percent of all roads and streets are paved with asphalt. Over 70 percent of the airport runways are paved with asphalt.

Other petroleum products include insect sprays, weed killers, rust preventatives, LP (liquified petroleum) gas, and petrochemi-

Asphalt is a petroleum product used to surface roads.

cals. Petrochemicals are chemicals made from petroleum. They are used to make such products as paint, detergents, synthetic rubber, plastics, antiseptics, cosmetics, drugs, anesthetics, fertilizers, and synthetic fibers such as nylon. Most of these modern petroleum products are made by *blending* several materials to produce the desired product. Over 3,000 different products are made from petroleum and natural gas.

Paper Making

Preprocessing paper. Loggers cut or harvest the trees by hand with axes, hand saws, or power chain saws. Trees harvested for pulpwood usually are cut into 4 ft. or 5 ft. lengths (*reducing in size*). Some modern pulp mills can now handle full-length logs. This eliminates the need to cut the logs into shorter lengths. Some trees are cut with large cutting machines. Some of these have a saw blade mounted on an arm that can reach out to a tree and cut it. Others have a large claw-like cutter that "snips" off the tree. A clamping device holds the tree while being cut. It also is used to lift the log onto a pile or load it directly onto a truck.

Once cut, the logs must be transported to the paper mills. The logs are usually hauled out of the forests on logging trucks. If the

Pulpwood is cut and loaded onto railroad cars for the journey to the paper mill.

paper mill is nearby, the trucks haul the logs directly to it. If not, the logs are loaded onto railroad cars, barges, or ships to be hauled to the mills.

The logs are unloaded at the mills and piled in a storage yard. Huge cranes and other unloading machinery are used for this. Logs are stored at the mill to insure a continuous supply of wood to the mill. Sometimes the harvesting of logs is slowed by bad weather. When this happens, the mill does not have to slow down because logs can be taken from the storage yard.

Logs are carried into the mill on conveyor belts. Sometimes water flumes are used. Once in the mill, bark must be removed from the logs (separation). Bark does not contain much wood fiber and must be removed from the pulpwood logs. This bark is used for fuel and for mulch around trees and shrubs.

The individual wood fibers in the logs must now be separated so they can be reformed into sheets of paper. This is done either mechanically or chemically. In mechanical pulping huge, rough grinding stones grind the logs into fibers (reducing in size). The fibers are mixed with water (pulp) and formed into sheets of paper.

Chemical pulping is done in a digester, which works like a pressure cooker. The logs are cut into small chips by rotating knives (reducing in size). These chips are *mixed* with chemical solutions in the digester, and cooked (*conditioning* by heating). This cooking dissolves the material (called lignin) that holds the wood fibers together. The fibers then *separate* into a wet, pulpy mass.

right The debarked logs are ground into chips to separate the individual fibers in the wood. *below* Woodpulp can be mixed with bleaches, dyestuffs, and sizing.

Once the wood has been turned into pulp by either of the pulping methods, it must be further treated. Bleaching gives the pulp the whiteness needed for some papers. Washing is needed to remove any chemicals remaining from the pulping. Screening removes bits of dirt and pieces of undigested wood. Often the pulp must be screened to separate different sizes of fibers. These are separating processes.

The pulp is now ready to be made into paper. This is done in machines called Fourdrinier machines. (Named after Fourdrinier, a Frenchman who invented the process). In the beater or blender, various kinds of pulp are *mixed*. This gives the paper special qualities. Dyes are added (blending) when colored paper is wanted. Water is added until the mixture is about 99 percent water.

This mixture of wood pulp and water flows into the Jordan where the fibers are cut and rubbed to the desired length (reducing in size). This helps produce a smooth sheet of paper without lumps. It also helps produce a sheet with uniform thickness.

The diluted pulp flows onto the huge paper machine. It first flows onto a fine mesh screen which allows the water to drain out of the pulp (separating). This screen may have as many as 6,000 holes per square inch. In some machines suction is used under the screen to help pull the water from the pulp. The screen shakes

Processing Technology

and moves to cause the fibers to weave and mat together. The paper is now a soft soggy mass of fibers. It hardly looks like the sheet of paper it will soon become.

This sheet of pulp now moves onto a belt made of wool felt. This carries the pulp through a series of rollers that press out more water. The belt absorbs some of the water starting to dry the pulp. The sheet of paper now being formed is strong enough to support its own weight. It is lifted from the felt belt and moves through a series of steam-heated rollers called dryers. In some machines the dryer section is as much as 110 meters (350 feet) long. The paper may now be moving as fast as 47 kilometers (30 miles) per hour as it moves through the machine and is rolled into huge rolls. The dryers remove all but about 5 percent of the water.

Paper must go through other processes before it can be used. Some paper is coated with a clay coating to give it a smooth surface. Some is sized (coated with clay) so it will not absorb too much ink when used for printing. Other papers require a smooth surface produced by ironing the paper between heavy, polished steel rollers. Some papers have designs printed on them. These are combining processes in which materials are combined with the paper to give it special qualities.

Some machines produce a piece of paper as much as 7.5 meters (25 feet) wide. This, obviously, must be cut into smaller sizes (reducing in size). Often this is done automatically by slitters when the paper is rolled up as it comes from the machine.

The pulp, diluted with water, flows onto the fine-mesh screens of the paper machine. The fibers are matted together and proceed through a series of rollers which press out the water.

At the dry end of the paper machine, the paper is wound into huge rolls.

Postprocessing paper. Once the paper comes off the papermaking machine it is stored. Often it is stored in the rolls that came off the machine. When a special paper or size of paper is needed, these rolls are placed on special cutters, which cut it into large sheets. These are then cut into the smaller sizes needed by the customer. The warehouse at the mill often contains rolls of many types of paper and cartons of cut papers waiting shipment to distributors, who sell it to the customers. Paper used to make cardboard boxes, bags, and other packaging is often stored in rolls. These rolls of paper are then placed on the box or bag-making machines. Finished papers are transported by truck and railroads to the many distributors across the country.

Paper must be cut according to the needs of the customer, and then shipped.

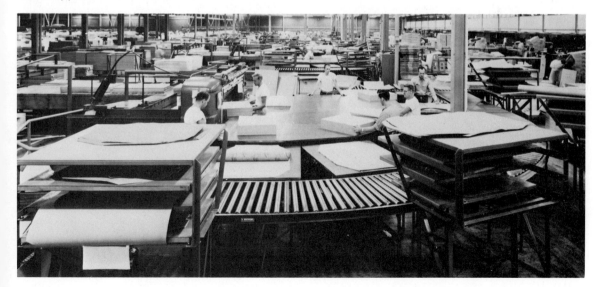

Processing Technology

Aluminum, extracted from bauxite ore, is cast into enormous ingots before being transferred to the rolling mills.

Processing Aluminum

The processing of aluminum provides another good example of processing technology. The methods used here are similar to the methods used in processing many of the nonferrous (non-iron) metals.

Preprocessing aluminum. Aluminum is the most abundant metal on the earth. Bauxite, the mineral from which aluminum is extracted, has been found on every continent except Antarctica. There is a little bit of aluminum in almost all areas of earth. But the methods of extraction used today require that bauxite ore contain at least 45 percent or more aluminum before processing is practical. Ore in this concentration is found in a wide band through the tropical and subtropical regions of the earth. This is a band extending on both sides of the equator. It is estimated that there is enough aluminum in the earth to supply our needs for thousands of years.

Bauxite ore usually is found in shallow deposits close to the surface of the earth. This allows it to be mined using surface methods. Strip-mining operations using power shovels and drag lines are commonly used to remove the ore. (See earlier description of surface mining.)

Processing aluminum. Once mined, the bauxite ore is transported to the aluminum plant, usually in trucks or train cars. Here it is crushed (reduced in size), washed, and dried. The dried and crushed bauxite ore is mixed with crushed lime and soda ash and then mixed with water (combining). The lime and soda ash react with the impurities in the ore. The mixture is allowed to stand in large tanks to allow the impurities to settle to the bottom. These settled impurities are called "red mud." The liquid in the top of the settling tank contains the aluminum-carrying compounds called "green liquor." This is drained and pumped into precipitator towers. In these towers small amounts of alumina crystals are added to the "green liquor." This causes more alumina crystals to form. These new crystals are washed to remove any remaining impurities and then dried in roasting kilns. The temperature in these kilns reach 930° C (2000° F). This high temperature evaporates any remaining moisture.

Out of the kilns comes a fine white powder, which is about one half aluminum and one half oxygen. Before pure aluminum can be obtained, this oxygen must be removed.

The alumina powder is placed in smelting pots. These are rectangular steel containers lined with carbon. A carbon electrode is lowered into the alumina and an electrical charge put through the powder. This generates a high temperature and the alumina melts. The oxygen in the alumina joins the carbon from the electrode to form carbon dioxide. The carbon dioxide is drawn off, leaving pure aluminum in the pot. The molten aluminum collects in a layer in the bottom of the pot. It is drawn into crucibles and more alumina is added to the pot to continue the process.

The aluminum produced by this process is about 99.5 percent pure. Once processed in this manner, the aluminum can be shaped into thousands of products. It can be forged into machine parts, drawn into wire, or rolled into plates, sheet, or foil. It can be machine processed into lightweight, intricate parts for millions of appliances, machines, and electronic parts.

Rolls of lightweight aluminum wire and cable are used in the electronics industry.

Processing Technology

Working in Processing

Careers in processing technology are concerned with the location, extraction, and processing of raw materials. In forestry the work is with the living, growing tree and its use. Careers include forest management, timber production, fire protection, and land management. People are needed to cut the timber, transport it to the mills, and cut it into lumber, or make paper. Foresters plan forest lands and plant and cultivate trees. Some people working in forestry help plan and develop recreational areas such as parks and campsites.

Careers in the petroleum industries are concerned with the location, extraction, and processing of the oil. The processing of gas is a related area, and the work is similar to that in the petroleum industries.

The mining and processing of minerals offers a different kind of work. Miners must remove the ore from the ground. Some ores, particularly coal, are mined underground. This work is hard and often dangerous. Miners must learn to use big equipment that cuts the coal and carries it to the processing plants.

As in all the areas of technology, there are many types of careers. Research and development work requires trained scientists. The research and development scientists work to develop new materials. They also develop ways to make better use of raw materials. The pollution of our environment worries many people. Today many scientists are researching ways to eliminate this problem.

Aluminum chips are unloaded into hoppers at a smelter.

left Textile worker *right* Chemist.

The professional careers in processing require a college education. Engineers, economists, and scientists are needed in many areas. But not all work in technology requires college education. No area of technology can be complete without the many people who assist the scientists. These people have an important role in technology. Often each scientist needs several assistants. Work is challenging and often hard. It is also interesting.

Many people are needed to run the machines used to find, extract, and process raw materials. People must run the plants that process the materials. Workers are needed in the storage areas and for transportation of both raw materials and the materials after processing.

Work in the processing industries is often hard and requires physical stamina. The underground mines are often damp, dark, and noisy. Safety practices in the mines have made the work much safer than it once was. Surface mines are usually less hazardous than underground mines. Since mining operations are now done by machines, workers are required who are able to run machinery properly and safely.

Drilling for oil and gas is also a rugged, outdoor job. Work shifts are long and often must continue even in bad weather. The wells are often drilled in remote areas. The workers have to move when the well sites are moved. This can happen often.

The employment outlook in the processing industries is good. The demand for the raw materials needed by industry is increasing. The need for fossil fuels, coal, oil, and gas, is increasing, also. New resources are needed as those now known are being used. Coal will play a larger role as a fossil fuel as oil and gas become harder to find and to market.

There are several ways you can learn more about careers in processing. Get literature from the many processing industries. Many of the larger industries have brochures that describe the type of work that is done in their industry. Some describe salaries and working conditions. Go to a local school or community library. Find a copy of the *Occupational Outlook Handbook*. This book gives detailed descriptions of many occupations. It also gives estimates of the number of workers needed in each of the occupations. This is important to you. You may not want to consider working in a job that does not have much of a future. You may want to look for work in an occupation that has a shortage of workers.

Suggestions for Further Study

OBSERVATIONS

You can become involved in processing technology in many ways. This chapter has been written only as an introduction to processing technology. You can study the many areas of processing in more depth, for example, by looking into the environmental impact of one of the processing industries—petroleum, steelmaking, or coal mining. You can measure the amount of water pollution or air pollution that comes from processing industries. Many studies can be done very simply. Only your nose is needed to detect smelly fumes coming from a processing plant. You can see smoke floating into the air. You can see some water pollutants.

Your studies can include model making. Make a model of an oil refinery or of a steel mill. Make a model that shows the parts of an oil drilling rig. Models can be used in displays and shows in your school or community.

Many hobbies are related to processing technology. Weaving is one. You can spin thread from flax, cotton, or wool fibers. These can be woven into cloth. Many people like to cook. Cooking is part of food processing. You can grow your own vegetables or fruits (a renewable resource) and can, freeze, or make jelly from them. If you are a collector, rock and mineral collections are interesting. Have you ever seen a collection of different kinds of wood? These can be very beautiful besides being fun to collect.

Visit historical places in your community or state. Many are related to processing. Many of these show early ways of doing things. It is interesting to learn how your grandparents did things. The level of technology fifty years ago was quite different from what it is today.

Questions
1. What is processing?
2. Name the three stages of processing. What is meant by each?
3. Name several of our nonrenewable resources. Why is conservation of these important?
4. What are renewable resources?
5. Describe the major underground mining methods. When is each used?
6. Why is strip mining used? What are some of the problems with strip mining?
7. What are some of the most common processing methods?
8. Select a raw material and describe how it is processed into usable form.
9. Can you think of any materials used just as they are found in nature?

Key Words
processing
preprocessing
postprocessing
extraction
harvesting
raw materials
bauxite
renewable resources
nonrenewable resources
petroleum
recycling
solid waste
mining
smelting
contour mining
placer miner

dredging
open pit mining
strip mining
quarrying
shaft mines
drift mines
rotary drilling
cable tool drilling
refining
pulp
reducing in size
separating
combining
grouping or classifying
conditioning
farming

Recycling Aluminum

Concept: *Recycling*

Large quantities of materials are discarded every day. Many of these can be recycled to be used again. Recycling reduces the amount of raw materials needed from the natural supply.

When aluminum is recycled, bauxite ore is saved. Also, much less energy is needed to recycle aluminum than to produce aluminum from raw bauxite ore. Recycling aluminum cans is a good way to save raw materials and energy.

Many people are recycling aluminum cans as a way of making extra money. Recycling companies pay for scrap aluminum as well as other materials. Is there a can collection center or recycler in your community? Find the price being paid for scrap aluminum. By saving aluminum cans for recycling you may be able to make some extra money for your class, club, or for yourself.

Objective

In this activity you will make a simple can smasher. You can use it to flatten aluminum cans before taking them to the recycling plant. The smashed cans will take up much less space.

Procedure

This smasher can be made from materials available in any lumber store. Scrap lumber can be used to reduce the cost of the smasher. Using the scraps will also be a small effort in recycling.

The base of the smasher is made from $2 \times 8"$ construction lumber. The sub-base can be made from $2 \times 6"$ construction lumber. The handle can be made from $2 \times 2"$ lumber.

Make the handle any convenient length. About $36"$ may be a good length to try. Round off the edges. Shape and smooth one end of the handle to give a comfortable gripping surface.

Assemble the handle to the sub-base. Cut a square hole through the sub-base. This hole should be cut so the end of the handle can fit into it. Drill a hole in the sub-base so a screw can be put in to hold the handle in place. Drill holes for wood screws to hold the sub-base to the base.

Now cut the base and fasten it to the sub-base and handle sub-assembly. A metal plate can be added to the bottom of the base to make it stronger. Without the metal plate, smashing cans will chip away the base.

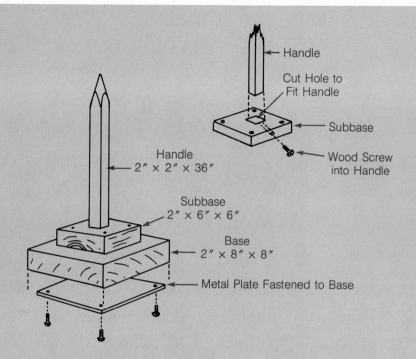

The hydroelectric generating plants at Hoover Dam supply energy for a large area of Arizona and Nevada.

Chapter 3   Energy and Power

Introduction

Technological progress depends upon energy and power. Energy is defined as the ability to do work. Power is defined as the amount of work done in a given amount of time. The power output of your body is determined by the energy contained in your muscles. This, of course, is determined by the amount of energy in the food you eat.

Early Stone age people were very limited in what they could do because they had only their muscles as sources of energy. They could only do those things they were strong enough to do. Slowly it was discovered that animals could be tamed and used as sources of energy. The animals were stronger. They had the ability to do more work than people. They were a source of greater energy.

This use of animals was a great advancement in technology. In fact, there was no other practical progress in energy and power technology for thousands of years. Until the nineteenth century, the horse was the fastest means of overland transportation and teams of horses provided power in large amounts.

As many as 4000 years ago it was known that sails could be put on ships to help move them along. In this way the wind was used only when the wind was blowing in the direction the ship wanted to go. Later it was learned how to sail across and into the wind. It was the power of the wind in the sails of ships that made possible the discovery of America by Europeans.

This waterwheel provided power to operate woodworking machines within the mill.

Other sources of natural energy were also used in early history. The Romans used the power of falling water or moving water in rivers to turn water wheels. Water wheels were used to grind grain and to power machines in colonial America.

The Industrial Revolution is one of the best examples of the effect sources of energy and power have on technological progress. During this period in our history, industry grew rapidly. This growth was made possible by adequate sources of energy. The steam engine provided power to run machines in factories.

Progress in farming is another example of the effect of energy and power on technological advancement. Early farming methods used the energy and power of the human muscle. Farmers harvested and threshed grain by hand. This was slow, exhausting

A horse drawn harvester with one operator could harvest more grain than several people using hand tools.

58 Exploring Technology

work. By 1831 horses were used to pull machines that cut the grain, making the farmer's work much easier. Gradually tractors began to replace the horse as a source of power in farming.

Examples can also be found in modern space technology. The astronauts and cosmonauts could not travel into space without the rocket power that got them past the pull of earth's gravity. Astronauts Scott and Irwin used a lunar rover powered by batteries to help them move on the moon's surface. With this powered vehicle they were able to explore more of the moon's surface and do more work and research while on the moon.

The machines and tools used to build the engines and machines themselves also needed power. By adding power to tools and machines it was possible to do work faster and more easily. It was also possible to work more accurately. The accuracy of machine tools was important in the manufacture of precision machines.

To understand technology one must understand the ways energy and power are used to make technology run. You need to know how power is used in the various areas of technology: transportation, communication, construction, and production. One also needs to understand that all these areas are interrelated. This means that each area depends upon all the others. All areas of technology depend on energy and power.

Think for a minute about this. What would happen if suddenly time were turned back to an era before the development of our modern power systems. All that would be available would be what

With this tractor-powered combine, grain could be harvested much faster than with the earlier horse drawn machines.

Energy and Power

one could do with one's own muscles and the muscles of animals. What would your life be like? What would you have to do without? What would no longer be possible without these technological advances in energy and power?

In this chapter you will learn about energy and power and their importance to all of technology. You will learn the sources of energy, how this natural energy is converted to useful energy, and then how it is used to help people do work. You will study some of the problems caused by modern energy and power systems and their uses.

Energy Sources

In this study of energy technology it is important that we study where energy comes from—the sources of energy. Nearly all of the earth's energy comes from the sun. All plants and animals on the earth need the energy of the sun to live. Green plants store energy as food through a process called *photosynthesis.* During photosynthesis solar energy is changed into chemical energy by the cells in the plant. Animals eat these plants and thus take some of the energy of the sun. Animals that are not plant eaters get the sun's energy by eating animals that do eat plants.

Solar energy causes changes in the earth's weather. Winds are created. Moisture is evaporated from lakes, rivers, and oceans. Clouds are formed, and the water falls back to the earth as rain or snow. This water, which has been moved from low areas of the earth to the higher areas, can do work as it runs back down the rivers to the low areas.

The materials we burn as fuel—coal, natural gas, petroleum, wood—are all products of the sun's energy. Millions of years ago plants growing in dense forests and swamps died and fell to the earth. These dead plants became tangled masses of decayed vegetable matter. More plants grew, died, and fell on top of this mass. This continued for millions of years. Heat and pressure caused the decayed matter to form coal. Where the heat and pressure were greatest, hard coal was formed. The energy stored in coal is actually energy from the sun that has been stored for millions of years.

Petroleum has been formed in a similar manner. Oil and natural gas, therefore, are stored energy from the sun.

For most of the time human beings have existed, only human muscles were available as a source of useful energy. Early people were able to exist only in a primitive way. They gathered food, built shelters, traveled, and made simple tools and weapons. These

Sources of Natural Energy

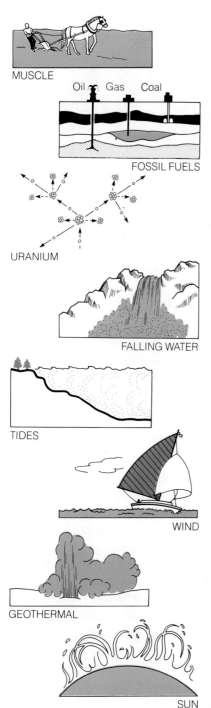

MUSCLE
FOSSIL FUELS
URANIUM
FALLING WATER
TIDES
WIND
GEOTHERMAL
SUN

early people were nomads—they traveled from place to place. They never stayed in one place for very long. This began to change when they began to learn to grow crops for food rather than merely gather wild foods. As they developed this ability to grow food, agriculture was born. It became possible to stay in one place and not move about constantly hunting food.

Agriculture brought about a big change in the lives of these early people. Because they did not have to move about as much, they were able to tame animals. These animals could be used as a source of energy as well as food. The muscles of the animals could be used to do work in place of human muscles. For the first time people were able to do work that required more power than the human being could provide.

Even with human energy, many remarkable things were accomplished. The great pyramids of Egypt, which are still standing, were built entirely by human labor. The Pharoahs used thousands of slaves, many years, and some rather ingenious methods to cut the stones, move them to the work sites, and put them in place.

It was thousands of years before the energy supplied by the muscles of man and animals would be replaced with other energy sources. The speed of transportation was limited to the speed of a fast horse. Such animals as oxen, camels, horses, dogs, and mules were used to carry loads. The invention of the wheel in prehistoric times made the tasks of the animals, including human beings, easier, but they still had to provide the energy.

The Egyptians used cattle and other animals to pull plows. It has been estimated that in the United States in 1918 most of the more than 22 million horses and mules were being used in agriculture. By 1950 this number had been reduced to 7½ million. The tractor replaced the horse and mule on the farm. Can you imagine how much land had to be used just to grow food for these 22 million horses?

As you can see, human muscles and animal muscles were our most important sources of energy for centuries. In a great many places on the earth this is still true today. But, it is important to understand more modern sources of energy to be able to live intelligently in our technological society. During this study you will also discover that some of the early sources of energy may become sources of the future.

Here we will study the following sources of energy: fossil fuels, nuclear, water, wind, solar, and geothermal. All are in use today, some more than others.

Energy and Power

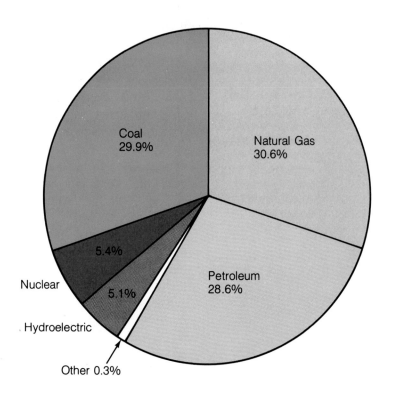

Source: Annual Energy Review 1985

Fossil Fuels

Coal, natural gas, and oil make up the fossil fuels. These are the most important source of energy today. In 1984 fossil fuels made up about 92 percent of all the energy used in the United States. Oil produces about 29 percent of this, gas 30 percent and coal 30 percent.

Hydroelectric and nuclear sources provide most of the remaining energy used in the United States. The burning of wood for heat has had some effect on these figures. It is difficult to measure the amount of energy produced from this source. Some estimates are as high as 3% of the total energy consumed. These percentages are changing as reserves of fossil fuels are used up. We are burning our fossil fuels at a very fast rate. Some scientists estimate that at present rates of use, coal reserves will last for several hundred years (some estimate as many as 600 years); oil for only thirty

HISTORICAL DEVELOPMENT OF ENERGY AND POWER

Time	Development
	Human muscle
	Animal muscle
400-200 B.C.	Waterwheel
400 A.D.	Falling water used as energy source
600-1000	Windmill used
1000	Water wheels drive mills
1673	Powder engine
1675	Spiral spring for watches
1707	High pressure boiler
1775	Water turbine
1775	Watt's steam engine
1782	Double-acting rotary steam engine
1787	Fitch's steam boat
1800	Trevithick's light-pressure steam engine
1800	Volta makes first battery—electricity from a cell
1801	Trevithick's steam train—first steam powered passenger vehicle
1803	Fulton's steamboat
1814	First practical steam engine
1819	Oersted discovers electromagnetism
1821	Faraday discovers fundamentals of electromagnetic rotation
1829	Henry makes early version of electromagnetic motor
1859	First oil well drilled
1860	Lenoir makes first practical internal combustion engine
1880	Winhurst makes electrostatic generator

continued

years; and natural gas for only twenty years. Others say these fossil fuels will last much longer. But, the fact is that the fossil fuels will not last forever. They are nonrenewable resources (see Chapter 2 for a discussion of nonrenewable resources.)

Scientists and technologists are working hard to find more fossil fuels, to find ways to conserve the fossil fuels we now have, and to find other materials that can be used in place of the fossil fuels. This is slow work.

Nuclear Energy

Many scientists and technologists believe nuclear energy can and will replace fossil fuels as our main source of energy. But there are many problems with nuclear energy. Most importantly, it is very dangerous. It must be controlled very carefully to prevent exposing people to radiation, or invisible rays of powerful energy. The wastes from a nuclear generating plant must be stored for hundreds of years before they are safe for human contact. These generating plants also require large quantities of water for cooling. This water, when released from the plant, is hot and can have serious effects on the fish and plants in the rivers and oceans.

In the process of nuclear fission the nucleus or center of the atom of the fuel (uranium or plutonium) is split. This gives off radiant energy which produces heat. Neutrons or parts of the atom also are released. These collide with the center of more atoms causing them to split. As the reaction continues, intense heat is given off. This heat is used to change water into steam.

Fission Reaction

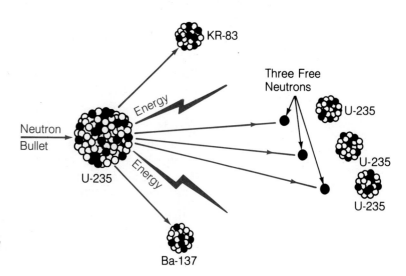

HISTORICAL DEVELOPMENT OF ENERGY AND POWER

Time	Development
1884	First practical steam turbine engine
1888	Tesla makes electric motor
1892	Diesel engine patented
1926	First liquid-fueled rocket
1929	Wankel rotary engine patented
1937	First jet engine
1942	Turbo prop engine
1951	Electricity produced from atomic energy
1954	Nuclear powered ship — Nautilus submarine
1958	Solar cells generate electricity for Vanguard I satellite
1959	Experimental fast breeder reactor — Scotland
1986	Chernobyl nuclear accident

The steam is used to drive turbines. The turbines turn generators that generate electricity.

If this chain reaction is not controlled, the uranium fuel can melt and pollute the surrounding environment with radiation. This happened in 1986 at the Chernobyl plant in the U.S.S.R. Of course, this possibility is a worry to all nuclear scientists and to people living near a nuclear power generating plant. But, scientists have learned to control nuclear fission and the chance of a nuclear accident is small.

After the fuel is used in a nuclear reactor, the waste must be disposed of. A certain amount of it can be reprocessed into more fuel that can be used in the power plants. The remaining waste still is radioactive and very dangerous. There seems to be no easy way to get rid of these materials. Burial in old salt mines has been tried but was not successful. Research into solving the problems of disposal of nuclear wastes continues.

Water

The water of the earth is another source of energy. This source also is affected by the sun. Energy from the sun evaporates water from the rivers, lakes, and oceans of the earth to form clouds. Eventually much of this moisture falls back to earth. Some water enters streams and rivers where it begins to run toward the oceans. The energy in this falling water can be trapped and used.

The old grist mills once in use in this country used the energy in flowing or falling water. Grain was ground into flour by these mills. Some of the mills also had woodworking or metalworking tools that were powered by the energy from the falling water.

Today the energy in flowing or falling water is used to turn generators and produce electricity. This is called hydroelectric generation.

For more than 2,000 years technologists have known that water flowing under or over a wheel could make that wheel turn. The undershot wheel was one of the first engines to do work. The paddles of the waterwheel dipped into the moving stream. The movement of the water created pressure on the paddles and caused the wheel to turn.

During the 1750s, John Smeaton, an English engineer, experimented with water wheels. He discovered that the overshot wheel was more efficient. Water was directed onto the top of the water wheel. Both the weight of the water in the wheel and the pressure of the water against the paddles caused the wheel to turn.

Exploring Technology

UNDERSHOT   OVERSHOT

These early waterwheels were the forerunners of our modern hydroelectric power plants. Operation of these plants will be discussed later in this chapter.

Wind

The wind, like flowing water, is a source of energy. Also like water, the wind has been used for energy for hundreds of years. The windmill is believed to have been in use as early as 643 AD. A windmill called the post mill was introduced into Europe in the twelfth century. It could be turned so it faced into the wind, no matter which direction it blew. In 1784 a fantail gear was added to the windmill so it would automatically rotate the head into the wind. When the wind struck the fantail, the head of the mill was turned into the wind. This worked very much like the rudder on an airplane. These early windmills were as much as 120 feet high. The mills had the main shaft connected by gears or belts to grinding stones, water pumps, or other devices.

The metal-vaned windmill was common on farms in the United States until recently. These mills were used to pump water from wells for cattle or to supply water for farm use.

But, windmills were limited. They only worked when the wind was blowing. If windmills are to be used as a major supplier of energy, then some method of storing the energy they produce is needed. At best, wind power will probably be used only as a supplementary source.

Solar Energy

The sun has been mentioned several times as the source of our energy. The examples discussed so far have all been of sources coming indirectly from the sun. The sun's rays are also a direct source of energy.

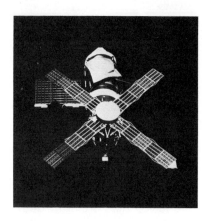

Panels of solar cells on the Skylab II Space Station convert solar energy to electrical energy.

Though the sun has always been there, early people did not know that they could use the sun for anything but light. They did not know that the sun gives off enormous amounts of energy. Much of this energy can be used. Scientists have been trying for years to use the sun's energy. Scientists estimate that only 13 percent of the sun's energy reaches the earth. Even this small amount of energy is useful.

It seems logical that we should try to make use of as much of this energy as possible. Solar energy seems to be nonpolluting. It is there every day, but solar energy is not available during the night. Less is available on cloudy days. Clouds block much of the sun's radiant energy. It is necessary to store the energy collected when the sun is shining so it can be used during night and during times when the sun is not shining.

Tidal Energy

Movement of the ocean tides can be a source of energy. Ocean tides move toward the shore, and twelve hours later move away from the shore. This is a movement of water similar to the movement of water in rivers. It is possible to place a sort of paddle wheel in the water and the moving tides will move the wheel in a similar manner to a waterwheel on grist mills. When the tide is coming in, the wheel will be moved in one direction. When the tide is moving out, the wheel will be turned in the opposite direction.

But, this description is too simple. Actually, a waterwheel, like those used on the grist mills, will not work very well. But, if the moving tide can be forced through a narrow opening like a pipe, then this narrow stream of water can be used to turn a turbine. (See the illustrations on pages 72 and 73.)

There are only a few places on the earth where the movement of the tides is great enough to make tidal energy practical. Tidal-powered electricity generating plants are working successfully in France. The United States and Canada have considered building a similar plant in the area of the Bay of Fundy near Nova Scotia. This is the only place in North America where the movement of the tides is great enough.

Geothermal Energy

The heat of the inside of the earth is a possible source of energy. As early as 1904 there was a steam generating plant built in Italy near Florence. Work on a geothermal steam generating plant began in 1960 in California. Since 1975 the Pacific Gas and Electric Company, in San Francisco, has been generating electricity using geothermal steam.

Geothermal energy is limited to those areas of the earth where steam and hot water are near the surface of the earth. When this occurs, wells can be drilled and the steam or hot water can be piped into turbines and electricity can be generated.

Scientists believe that very hot rock lies two to five miles below the surface of the earth in many places. In these places, it is felt that water can be forced down wells into the hot rock. The rock would heat the water into steam and this steam could be piped back to the surface where it could be used to turn turbines. The steam would be cooled and condensed into water which would be repumped into the wells to be reheated and reused.

There are many problems with geothermal heat as a source of energy. Though it is working successfully in some places, it is not practical everywhere and may not be practical on a large scale.

In this generating plant at the Geysers in California, electricity is produced from geothermal steam.

Future Energy Sources

Scientists are working on many other sources of energy. Some are only theoretical. This means that it seems they should work, but no one has proven they will.

Some feel that trash can be burned for heat energy. This heat then could be used to make steam, which could be used to turn turbines to generate electricity. Each year in the United States, there are tons of trash that could be burned. It is estimated that two tons of trash has the heat value of one ton of coal.

Energy Conversion

Nearly 25 percent of all the energy used in the United States is converted into electrical energy. This then is used in a variety of ways. First it is important to understand how the electricity is made or generated.

Most electrical energy is produced by a machine called an electrical generator. A generator converts mechanical energy (movement of a machine) into electrical energy (electricity). A generator must be driven by some outside source of power, usually some kind of turbine. The turbine or engine provides the mechanical energy necessary. In some cases, a portable generator is powered by a gasoline or diesel engine. The automobile generator and bicycle generator are good examples. In the automobile, the generator is powered by the engine; a bicycle generator is powered by the rider pushing on the pedals.

How Does a Generator Work?

Michael Faraday, an English physicist, discovered that by moving a copper wire through a magnetic field (an area of magnetic force surrounding a magnet) electricity would flow through the wire. This is called electromagnetic induction. If the wire is a part of a circuit or a loop of wire, then electricity will flow through the loop. As one side of the loop of wire moves through the magnetic field, the other side of the loop moves in the opposite direction. This causes the electricity to flow in one direction through the loop.

As the rotating loop continues to move, it moves parallel to the magnetic field of force from the magnet. At this point no electricity is generated. In the second half of the turn, electricity again is generated or induced. This time it moves in the opposite direction. Again the loop moves parallel to the magnetic lines of force and no electricity is generated. The loop continues to move through the magnetic field of force as the generator turns. This produces

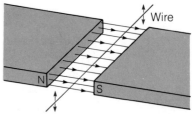

Electromagnetic Induction

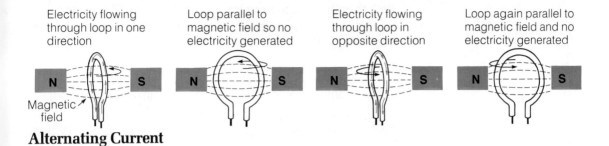

Alternating Current

what is known as alternating current because the electricity is changing directions.

What is a turbine?

A *turbine* is a wheel that is turned by the force of a liquid or gas striking blades on the edge of the wheel. This force can be air, water, or steam. The turbine is an energy converter. It converts the energy in the moving fluid or gas into mechanical energy.

Turbines are used to supply mechanical energy to electrical generators. These turbines are much more complex than the water wheel, windmill, or the pinwheel.

Steam turbines. Steam is used to turn the turbines in electrical generating plants. The steam is produced in a boiler and piped into the turbine. When water is heated into steam, the steam can expand to take up as much as sixteen times the space of the water. This causes a great pressure to build in the boiler. This high pressure steam then is directed into the blades of the turbine. This causes the turbine to rotate. After the steam leaves the turbine, it cools and condenses into water. This water is returned to the boiler where it is heated again into steam to go back to the turbine.

Water turbines. Water turbines are used in hydroelectric and tidal power plants. In hydroelectric plants, dams or waterfalls cause the water to fall from a great height. The falling water strikes the blades of the turbine. This causes the wheel to turn.

Water turbines have the advantage of not needing fuel to produce steam to turn the turbine. As long as the water flows, there is a source of energy. Water turbines also are nonpolluting because no smoke or fumes are given off.

Water turbines have the disadvantage of needing large amounts of falling water. Where there is no natural waterfall a

Energy and Power

dam must be built to raise the level of the water. This means that hundreds of acres of land must be flooded. Sometimes this is not desirable.

Gas turbines. Gas turbines are used to turn generators and as power plants in locomotives, ships, and airplanes. Gas turbines use the energy in hot gases to turn the turbine. A burning fuel produces these gases, which are forced against the blades of the turbine.

A jet engine is an example of a gas turbine. (See the section on jet engines.)

Steam Generating Plants

Nearly all electricity generated today is generated in steam turbine power plants. In these plants a heat source such as a burning fossil fuel heats water until it turns into steam. This steam, under high pressure, is directed against the blades of a turbine, causing it to turn. This turbine is connected to a generator which, when turned, produces electricity.

The fuel is burned in a firebox. The heat from the burning fuel heats water in a boiler. Here the hot gases from the burning fuel flow over tubes that are filled with water. As the steam forms, it flows to the tops of the tubes. It then flows through a superheater where it is heated more. This produces superheated steam under high pressure.

All steam generating plants work in this same way. A fuel is burned to produce heat, which converts water into steam. This steam is used to turn a turbine, which turns the generator. The major difference between the several types of steam generating plants is the type of fuel used: fossil fuels or nuclear fuels.

Steam turns a turbine wheel such as the one in this photograph. The turbine turns a generator which produces electricity.

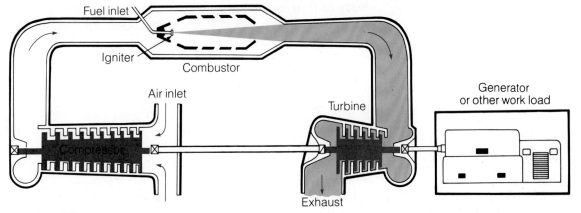
Gas Turbine

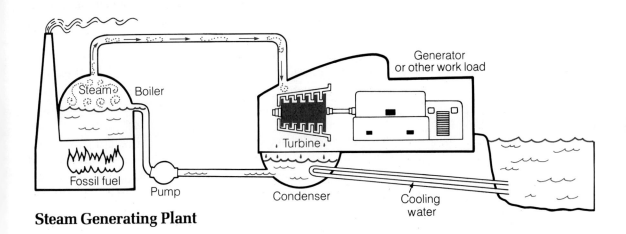
Steam Generating Plant

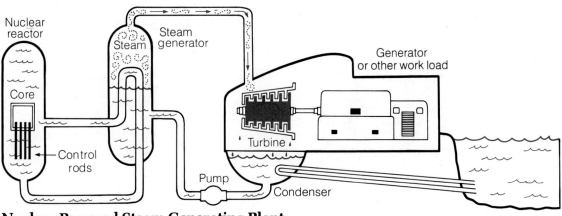
Nuclear Powered Steam Generating Plant

Energy and Power

Nuclear Power Plants

Nuclear power plants "burn" fuel such as uranium. There are two ways of producing atomic energy. In the fission method, the atoms of radioactive materials such as uranium are split into smaller, lighter atoms. In the fusion method lightweight atoms are united into larger atoms. In both methods atomic energy is released. A large amount of heat is also produced.

At the present time only fission is used to produce power. Atomic power plants and the atomic engines of ships and submarines use the fission method. Airplanes and space craft are also being planned that will use nuclear power.

In a nuclear power plant the heat produced by the nuclear reaction is used to convert water into steam. This steam is then used to turn turbines, which turn the generators.

The core of the nuclear reactor is the *firebox* of the nuclear power plant. This core is made up of fuel elements that contain uranium. The fuel is in the form of rods which can be pushed into the core and withdrawn to control the reaction. Water or liquid sodium is circulated through the core. This picks up some of the heat of the core and carries it to a heat exchanger. Here the heat is used to change water into steam. This steam is then piped into a turbine, which then turns the generator.

Hydroelectric Power

Water must be flowing from a higher elevation to a lower elevation before it can generate power. The force of gravity causes the water to fall, thus providing the source of energy. The kinetic

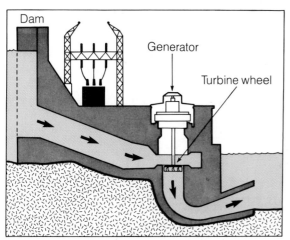

Hydroelectric Generating Plant

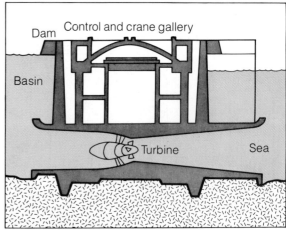

Tidal Power Plant

(moving) energy of the falling water is turned into mechanical energy, which is converted into electrical energy by the electrical generator.

The difference between the high level of water and its low level (such as the difference between the level of the water at the top of a dam and the level at the bottom) is called the *head* of the water. The larger the difference, the greater the head and the more potential energy in the water.

In hydroelectric power plants, this head is gained either by damming up the water or by using the natural head present in waterfalls. The greater the drop of water, the more force the water can exert. More power can be generated.

A turbine is placed in the falling water. The water pushes against the blades of the turbine and turns it. If a generator is connected to the shaft of the turbine, electricity can be generated. The mechanical energy in the turning turbine is converted to electrical energy by the generator.

Tidal Power

Tidal power works similarly to hydroelectric. The head of water is obtained from the difference between the high tide level and the low tide level. As in hydroelectric plants, water must fall or push against turbine blades to force them to turn. In tidal power plants, a damlike structure is placed across the mouth of a coastline bay of water. As the tide rises this dam forces the rising water to flow through a tunnel in which a turbine has been placed. The turbine is turned by the force of the incoming water.

As the tide begins to fall, the water flows outward through the tunnel where it turns the turbine again. The water flows through the turbine until the water level behind the dam nears the level of the ocean at low tide. At this point the turbine is stopped and the blades are reversed. When the tides begin to rise, the process begins again.

The use of the energy in the tides is not new. In 1734 in Chelsea, Massachusetts, spices were being ground at Slades Mill. The mill was powered by a four-wheel tidal unit. It is estimated that this mill generated as much as fifty horsepower. In Rhode Island there was a similar tidal-powered mill which had wheels that weighed 20 tons and measured 3.3 meters (11 feet) in diameter and 7.8 meters (26 feet) wide.

As many other early sources of energy, the use of tidal power declined when electricity became cheap. Small tidal-powered mills could not compete with the electric-powered mills.

Water rushing through the tunnel of a tidal generating plant turns the blades of this water turbine which in turn powers an electrical generator.

Tidal power is possible in only a few places on the earth. An extreme difference between the level of low and high tides is required. There are several possible sites in France, England, the Soviet Union, and South America. The only North American site is the Bay of Fundy, located between Nova Scotia and New Brunswick in Canada. There are nine possible sites in the bay where tidal power plants could be located.

The only site in the United States is Passamaquoddy Bay on the coast of Maine. Here the tides may rise as much as 15.2 meters (50 feet) above low tide. Harnessing the tides of this area has been discussed for many years.

Geothermal

Geothermal energy is energy from the heat from the inside of the earth. Volcanoes are a good example of geothermal energy. Geysers such as *Old Faithful* are created by geothermal energy. It may be possible to extract huge amounts of energy from the natural heat in the earth. If this source of energy can be tapped, it should last forever.

The use of geothermal energy is not new. In the 1930s Iceland first used geothermal hot water to heat houses. In the town of Reykajavik, 90 percent of the homes are heated with geothermal hot water. More than 100 wells supply the water. An elaborate system of pipes carry the water as far as 16 kilometers (10 miles). This hot water also is used to heat baths, swimming pools, and greenhouses. The greenhouses supply fresh vegetables year round. Russia, Japan, and New Zealand also make use of geothermal heat energy.

Electricity can be generated from geothermal steam. A large geothermal field is located at Larderello in Italy. Electricity has been generated there since the early 1900s. In California in an area known as *The Geysers*, an electric power generator plant uses geothermal steam. This is the only such plant in the United States. Over 100 wells have been drilled here. The deepest is about 2,300 meters (8,000 feet) deep.

The first step in developing geothermal energy sources is to locate the heat sources. Most geothermal sources in use today are located near hot springs. Other sources can be found using petroleum prospecting methods. (Refer to Chapter 2, Processing Technology.) Methods of drilling the wells are also similar.

Though it is being used in several parts of the world, geothermal energy is far from perfected. First, all areas are not the same. Some produce steam alone. These areas are called *dry steam* fields. Some areas yield mixtures of water and steam. These are called *wet steam* fields. A third area of possibly useful geothermal energy consists of hot rock. It has not come into contact with water and hence is dry.

The dry steam fields are relatively rare. The wet steam fields are about twenty times more plentiful. They also are more difficult to develop into working energy sources. The technology of exploring and using hot rock as a geothermal energy source is at a very early stage.

Once located, and brought to the surface, geothermal steam can be piped into steam turbine generators to generate electricity. In areas of dry steam wells, the steam can be used directly in the turbines. In wet steam areas or only hot rock, the technology is more difficult.

Wet steam wells contain steam and water mixed with numerous minerals. Often the steam can be separated to turn the turbines. The water then can be used for many different purposes. Minerals that are dissolved in the water cause many problems. They clog pipes and heating equipment. Often the water contains large amounts of salt. This salt must be removed before the water can be dumped on the surface of the earth. But, if separated out, the minerals can be put to use. This then can be a source of needed minerals as well.

Wind

The wind is moving air. It is caused by the heating and cooling action of the sun on the earth. Historically the wind has been used as a source of power. From ancient times sailing ships have been important in means of transportation. Sailing ships were used by the Egyptians as early as 2800 B.C.

Energy in the winds also was converted to mechanical energy by windmills. Windmills are known to have been used in Persia between the seventh and tenth centuries.

A windmill's operation is rather simple. Wind pushing on the blades or sails turns a shaft. This shaft is connected to a pump. The typical windmill could develop only low power. This limited the work it could do.

The most famous windmills are those used in Holland where over 30,000 were in use in the latter part of the nineteenth century. The Dutch used windmills to reclaim land from the sea. They built dams or dikes around an area of land. The sea water was then pumped out with the windmills. When the land dried, it was used for farming.

Windmills were used on a large scale during the nineteenth century in America. The settlers in the Great Plains in the United States needed a reliable source of energy to pump water. The regular winds of this area of the country made windmills a practical source of energy. Windmills could be a possible future source of energy. Research is under way.

Solar Energy

Direct solar energy can be used in many ways. Photovoltaic cells, called solar cells, change the sun's energy directly into electrical energy. A solar cell is made of materials similar to those used in transistors. Two of these materials are sandwiched together. Photons in sunlight strike the electrons in the materials of the cell causing them to be released. A number of these solar cells can be connected in an array to generate a useful voltage.

Heat from the sun can be collected in various kinds of collectors. A simple box put out in the sun will collect heat and get warm. This, of course, is not very practical because heat must be saved to be useful. Often it must be transported away from the collector to the place where it is needed.

Flat plate collectors have been developed to collect solar heat. Some medium such as air or water is used to carry the heat away from the collector and to store it for later use.

Forced-air collectors are often used to heat homes. Fans force the heated air from the collector into the house to warm it. Cool air from the house is circulated back into the collector to be heated. In hydronic collector systems, a liquid, usually water, is circulated by pumps through the collector. In areas where outside temperatures fall below freezing, antifreeze must be added to the water. As the water flows through radiators inside the house, the heat is given off to warm the building.

Heat is collected only when the sun is shining. On cloudy days little heat can be collected. At night no heat is collected. The heat must be stored so it can be used during the night or on cloudy days. In hydronic systems the heated water can be stored in tanks, which are often filled with rocks. The rocks, once warmed by the water will hold heat for a long time. When heat is needed, the water is pumped from the storage tank through the heating system in the house.

The heat collected in moving air solar heating systems can also be stored. Rocks, or other materials with a large mass, can be put in an airtight box. Warm air is pumped into the box to warm the rocks. When warm air is needed it is pumped out of the box into the house.

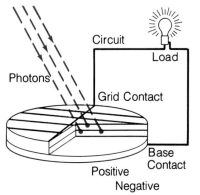

Photovoltaic Solar Cell

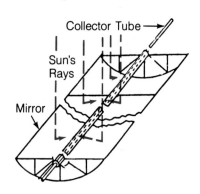

Simple Parabolic Trough Concentrator

Solar Air System

Solar Liquid System

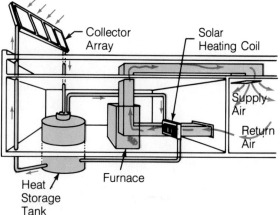

Most solar heating systems need a back-up heating system for those times when the solar heating system does not produce enough heat. In some areas there is not enough sunshine for a solar heating system to be practical. In other areas the winter temperatures may be low for long periods of time. The solar collectors may not be able to produce enough heat. Solar heating has the big disadvantage of being practical only in areas where there is a lot of sunshine. Often solar heating systems are added to regular furnaces. The furnace is used only when the solar heating system does not produce enough heat.

A simple flat plate collector looks like a box with a transparent (usually glass or plastic) top. Inside the box heat-collecting baffles are painted a dark color. The darker color absorbs more of the sun's energy. In a hydronic system pipes run through the baffles to carry the liquid. In air-type collectors, air is circulated through the baffles. The box is insulated to reduce heat loss.

In a passive solar system, no pumps or fans are used. Air or water moves through the system by natural convection and conduction. In a simple passive system, barrels are filled with water and painted black. These barrels are built into a south-facing wall of a house. During the day the barrels of water collect solar heat. During the cooler night the warm, water-filled barrels give off heat, warming the inside of the house. Stones, bricks, block or other masonry materials are also used as collector walls instead of barrels of water.

A simple passive solar collector can be put in a window. Air is heated by the sun, rises and flows into the room. Cool air is drawn into the collector to be heated.

Solar energy can be concentrated by parabolic reflectors. One type of parabolic reflector looks similar to a television satellite antenna dish. The parabolic dish reflects the sun's rays onto a collector suspended above the center of the dish. Temperatures near 1000 degrees fahrenheit have been attained in this type of solar collector.

A trough-type solar collector also concentrates the solar energy. The long trough reflects the sun's rays onto a long tube. Water can be run through the tube to be heated. Very high temperatures also can be attained in the parabolic trough collector.

Passive Solar Window Collector

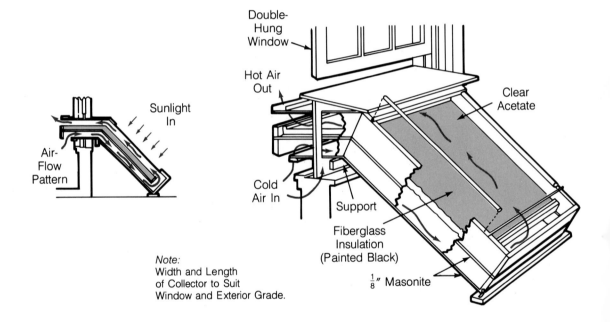

The Steam Engine

Though used little today, the steam engine was very important in the development of the technology of energy and power. It was the steam engine that helped make the industrial revolution possible. The steam engine was the major source of power for industry and transportation for many years, actually until the development of the internal combustion gasoline engine.

As early as 100 years before the birth of Christ, Hero of Alexandria made what many people consider to be a steam engine. It is actually as much an illustration of a jet engine as it is a steam engine. Hero's engine worked very similarly to a lawn sprinkler. The force of steam coming from the nozzles made the "engine" spin.

This was little more than a toy and did no useful work. It was not until early in the 1700s that a working steam engine was developed. Thomas Newcomen, an Englishman, developed a steam engine that could be used to pump water from coal mines. In those days, the depth of coal mines was limited because below certain depths, water would fill the mine shafts. For many years inventors tried to develop a pump that would remove this water.

Newcomen's engine used a large cylinder with a piston hung into it from the end of a seesaw-like beam. Steam was let into the cylinder. Cold water was then sprayed into the cylinder, which cooled the steam and caused the resulting vacuum to suck the piston down. This caused the other end of the beam to move up. That end of the beam was attached to the pump in the mine, and water was drawn from the mine.

In 1763, James Watt improved the Newcomen engine so much that many people credit Watt with inventing it. This is, of course, not true though his improvements were very important. Watt's improvements made the steam engine more efficient and less expensive to operate. His improvements made the steam engine practical enough so that it could be used for work other than pumping water out of mines.

Modern steam engines operate much like the engine developed by Watt. The major improvement has been the development of high-pressure steam engines. Watt's engine worked with about 15 pounds of pressure per square inch. This is only slightly greater than atmospheric pressure. In the early 1800s Richard Trevithick developed a steam engine that operated at nearly 30 pounds of pressure. Around 1815, Oliver Evans, an American, developed an engine that operated at 200 pounds of steam pressure. Modern steam engines operate with over 1,000 pounds of steam pressure.

Other modern improvements in the steam engine are the use of compound cylinders (two cylinders acting together), the use of super-heated steam (steam heated to over 370°C [700°F]), and improved valves for controlling the steam.

Equally important is the fact that efficient steam engines were not possible until precision machines were developed to make the

The *Trevithick,* a steam engine, was built in 1802 by Richard Trevithick.

engine parts. Machine tools were needed that could drill holes nearly perfectly round so steam would not escape. Efficient engines depended upon precision-made parts. This required the development of precision machine tools—another important development of the industrial revolution.

About three-quarters of all energy used is converted to forms other than electrical. Roughly one-third of this is converted into mechanical energy in the engines in automobiles, trucks, buses, airplanes, boats, ships, and other vehicles.

Internal Combustion Engines

Internal combustion engines are so named because fuel is burned inside the engine. This burning fuel provides the power necessary to move the vehicle. Gasoline engines, diesels, and rockets are internal combustion engines. All of these use petroleum products as their fuel.

Though internal combustion engines appear to be different, they are essentially the same. All require some chamber in which the fuel is ignited or burned. They all need fuel, and they all need air to be mixed with the fuel so it will burn.

Four-cycle gasoline engines. Most engines in use today are gasoline engines called *four-cycle engines.* This means that as the engine runs, it goes through a series of four stages or cycles: intake, compression, combustion and exhaust. These are repeated many times as the engines run.

Intake—the intake stroke starts with the piston at the top of a cylinder. As it moves downward, the intake valve opens, allowing the fuel to be drawn or pumped into the cylinder.

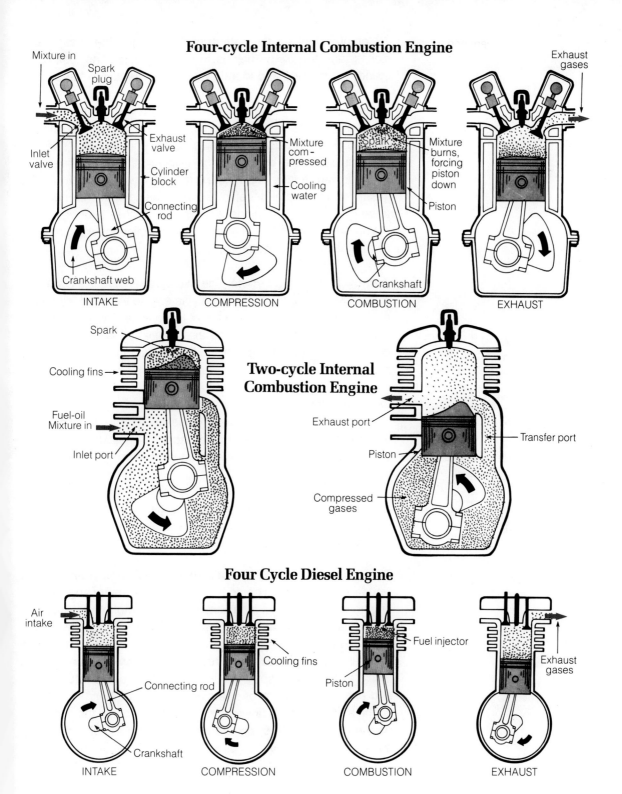

Compression—At the bottom of the cylinder, the intake valve closes, and the piston begins to move upward. The fuel in the cylinder is trapped and therefore is compressed into the smaller cylinder opening.

Combustion—At the top of the stroke, the fuel is ignited by the spark plug, causing the fuel to burn or explode. This drives the cylinder down, providing the power to the engine.

Exhaust—At the bottom of the cylinder, the piston begins to move upward again. Then the exhaust valve opens, allowing the gases from the burned fuel to escape, making space for new fuel to be drawn into the cylinder to start the process over again.

You will want to study the four cycle gasoline internal combustion engine more in the laboratory.

Two-cycle gasoline engine. The *two-cycle engine* is a commonly used small engine. In the two-cycle engine the spark plug fires each time the piston is at the top of the cylinder. When the piston is at the top of the cylinder, the air-fuel mixture is drawn into the crankcase through a reed valve. As the piston moves down, the reed valve is closed by the fuel mixture, which is being compressed. This prevents the air-fuel mixture from going back into the carburetor.

As the piston moves down, the exhaust port is uncovered allowing the exhaust gases to escape. The intake port is also uncovered, allowing the air-fuel mixture to enter from the crankcase.

As the piston once again moves upward, the air-fuel mixture is compressed. When the piston reaches the top of the cylinder, the spark plug ignites the mixture, causing the piston to be driven downward. This provides power to the engine.

Rotary Engines. In 1963 the Japanese NSU Spyder was the worlds' first production car to feature a Wankel rotary engine. This engine was developed by a German, Felix Wankel, in 1956. The Wankel is one of several rotary engines that have been developed.

The rotor and the drive shaft are the only main parts of the Wankel engine to move. The rotor is three-cornered and moves inside a chamber shaped to fit the movement of the rotor. Intake and exhaust ports are on the side of the chamber.

The air-fuel mixture flows from the intake port into the chamber. As the rotor turns, the mixture moves around the chamber until it reaches the combustion chamber. Here the spark plug ignites the mixture. The burning gases expand, forcing the rotor to con-

Wankel Rotary Engine

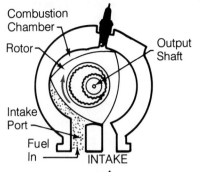
INTAKE

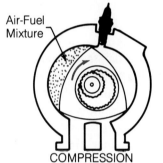
COMPRESSION

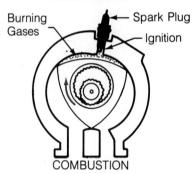
COMBUSTION

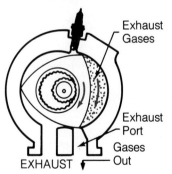
EXHAUST

tinue turning. The burned gases are carried around the chamber. When they reach the exhaust port the gases are released from the engine.

The rotor continues to turn. New air-fuel mixture is taken into the chamber, and the process continues. Notice in the diagram that the Wankel has what could be called three chambers. All three chambers are working at the same time. In the diagram fuel is shown moving through only one revolution of the engine.

Diesel engines. The diesel engine is similar in appearance to the gasoline engine. In fact, it works similarly; but, there are some very important differences. As this engine runs through one cycle, several of the major differences becomes evident. With the piston at the top of the cylinder, the intake stroke is begun, just as in the gasoline engine. But, here is the first difference. The intake valve opens and air is drawn into the cylinder. *No fuel* is drawn in. The piston continues to the bottom of the cylinder and moves upward, compressing the air in the cylinder. As the air in the cylinder is compressed, it gets hot. This happens more in the diesel engine because it operates at a higher compression ratio. In other words, the gases are compressed more.

When the piston is at the top of the cylinder, the diesel fuel is injected into the top of the cylinder. The heat of the compressed air causes it to ignite. This drives the cylinder downward, providing power to the engine. On the fourth stroke, the piston again moves upward, the exhaust valve opens, allowing the gases to escape. At the top of the stroke, the intake valve once again opens, allowing new air to enter, and the cycle begins again.

There is also a two-stroke diesel, just like the two-stroke gasoline engine. As in the gasoline engine, the two-stroke diesel engine fires each time the piston is in the top of the cylinder.

The two-stroke diesel engine uses a pump to force air into the cylinder when the piston is at the bottom of the power stroke. When the piston is in this down position, the exhaust ports open, allowing the burned gases to exhaust. At this point new air is pumped into the bottom of the cylinder, which also helps force the exhaust gases out.

As the piston moves upward, the air ports are covered and the air is compressed. The exhaust valves are also closed. When the piston reaches the top of the stroke, the fuel injector injects fuel into the top of the cylinder. The fuel ignites explosively, driving the piston down. This is the power stroke. As the piston nears the bottom of the cylinder the exhaust ports open, allowing the gases to escape, the air intake ports are uncovered and air is blown into the cylinder to continue the cycle.

Jet engines. Since the mid-1940s the jet engine has become an important part of our lives. Commercial and military aircraft, and some business and private aircraft, are powered by jet engines. This has resulted in faster flights. In many cases it is now possible to fly nonstop in a jet-powered airplane when a propeller driven airplane would have to stop to refuel. Air travelers now have more comfortable air travel. But, many of these travelers do not understand the essential principles of the jet engines on the airplanes in which they ride.

This section explains the jet engine. The differences between the three basic types of jet engine are also shown.

A balloon is a good illustration of jet propulsion. Blow up a balloon. If you quickly release the stem, it shoots around the room. What makes the balloon move? It is *not* the escaping air pushing against the air outside the balloon. In fact, the air outside the balloon tends to slow it! When the balloon was inflated the air pressure inside it was equal in all directions.

When released, escaping air releases the inside pressure. This is the same as the exhaust of a jet engine. On the side opposite the stem of the balloon, the air continues to push on the skin of the balloon. It is the push on this side that causes the balloon to move. This is called a reaction to the action of the escaping air.

The balloon does not move for very long. The air pressure is lost quickly. This could be overcome, if there were ways to put air pressure in the balloon. This would keep the jet of air moving out of the stem (exhaust).

Jet Propulsion

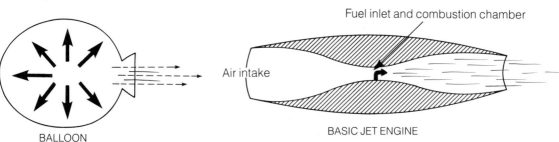

BALLOON

BASIC JET ENGINE

This is done in the jet engine by a compressor. The compressor is a spinning, fan-like motor in the intake of the engine. It runs at high speed and pumps huge amounts of air into the engine. One can see that this would be hard to put into a balloon. But, there must also be some source of power for this compressor.

Behind the compressor there is a combustion chamber. Here fuel is sprayed into the stream of air moving through the engine. This air-fuel mixture is ignited and it burns explosively. Since the compressor is forcing air into the front of the engine, the burning gases must exhaust out the back.

Turbojet. A turbine is placed in the exhaust stream. This is connected by a shaft to the compressor. So, the energy in the burning gases is converted into mechanical energy to drive the compressor. This keeps a supply of air moving into the engine. Energy in the burning gases also is expanded out the exhaust. The reaction to this action (like the reaction of the balloon) gives the forward thrust of the engine. The greater the amount of speed of the gases from the jet nozzle, the greater the thrust of the engine will be.

This is a turbojet engine. It is one of the simplest jets. Compared to propeller and piston airplane engines, the turbojet is lighter, smaller, faster, and more powerful. But, it burns more fuel at lower speeds.

Turbojets can be made more powerful by adding an *afterburner.* The thrust can be almost doubled. The afterburner works by spraying fuel into the jet exhaust. This added fuel ignites and burns explosively, increasing the amount of exhaust. The thrust of the engine is increased, thus increasing the speed of the aircraft. It has the advantage of added speed but the disadvantage of burning more fuel, particularly at lower speeds.

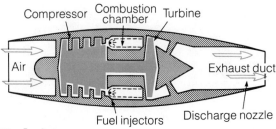

Turbojet

Energy and Power

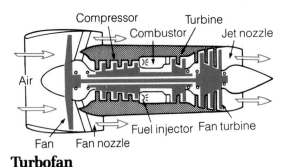
Turbofan

Turbofan. A Turbofan jet engine works much the same as the turbojet. In the turbofan a "fan" is added in front of the compressor. This causes the air to move faster as it goes into the compressor. The fan is connected to the shaft of the compressor and turbine. It gets its power from the turbine.

Some of the air that passes through the fan does not pass through the compressor. Instead it goes around the outside of the combustion chamber and turbine. It enters the exhaust near the rear of the engine. Because this air does not pass through the combustion chamber, it is not heated. By entering the jet exhaust, it still adds to the thrust of the engine.

The turbofan engine is more efficient, makes less noise, and has greater thrust than the turbojet. There is more power for

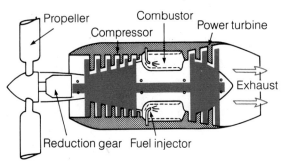
Turboprop

Exploring Technology

Simplified Rocket Propulsion Systems

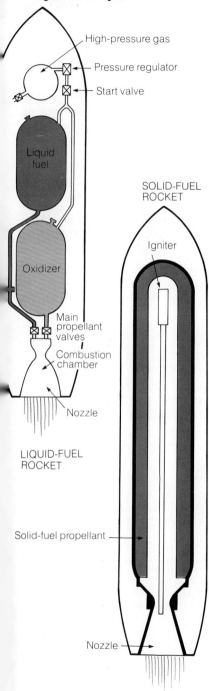

takeoffs and climbing. Because of its high thrust at low speeds, airplanes with turbofan engines do not need long runways for take off. They can also climb much faster. Many commercial airplanes today are powered by turbofan engines.

Look at the diagram of the turbofan engine. You can see that it is basically a turboprop engine. It has a large cowling around the fan. The prop has been changed into a many-bladed fan.

Turboprop. The third type of jet engine, the turboprop, is simply a turbine engine with a propeller attached on the front. The gases of exhaust provide some thrust but only a small part of the total thrust of the engine. Most of the thrust comes from the action of the propeller.

The propeller is connected to the turbine shaft through a gear box. This changes the high speed of the turbine to a slower speed needed for highest efficiency of the propeller.

The turboprop engine is used in smaller aircraft. It is best suited for slower speed aircraft. It is used in business and private aircraft. It is much quieter than piston engines with propellers.

Rockets. Rocket engines operate similarly to jets. The power for their forward thrust is provided by reaction to the exhaust of burning fuel. The big difference is that rockets carry their own oxygen supply. They do not need an air inlet. This allows them to operate in outer space where there is little or no oxygen.

Fuel and oxygen are carried in the rocket. These are mixed and ignited in the combustion chamber. The burning air-fuel mixture expands and forces its way out the rear exhaust of the engine. This provides the thrust needed to propel the rocket forward. Some rockets contain a solid fuel propellant, which contains the fuel and oxidizer (oxygen). Once this is ignited, it usually burns until all the propellant is burned. It is difficult to stop the burning of propellant. Once stopped, it cannot be restarted.

The solid fuel rocket can be stored for long periods of time. It can be ready for instant use after long storage. For this reason many military rockets use solid fuel propellant.

Liquid fueled rockets carry fuel and oxidizer in separate tanks. The two liquids are fed through a complex system of pipes and valves to the combustion chamber. Here they are mixed and ignited. Some fuel-oxidizer mixtures ignite when they come in contact with each other. Others, more commonly used today, require a spark or some other means of ignition.

Energy and Power

Liquid fueled rocket engines will continue to burn as long as the fuel and oxidizer come in contact with each other. The engine can be stopped simply by stopping the flow of fuel and oxidizer into the combustion chamber. It can be restarted easily. The burning of the fuel can be controlled by simply opening or shutting valves.

Transmitting Energy and Power

Transmitting Electricity

Once generated, electricity must be transmitted to the places where it is to be used. This is done through transmission lines. These lines are made of metals that conduct electricity easily. Silver is one of the best conductors of electricity. Of course, the cost of silver is too high for use in transmission lines. Copper and aluminum are good conductors and cost much less than silver. For this reason they are used to make the wire in transmission lines. Aluminum is not as good a conductor as copper, but it is much lighter. Where aluminum wires are used, larger wires can be used and the poles can be placed further apart. This helps make the aluminum wires more economical.

An Electrical Distribution System

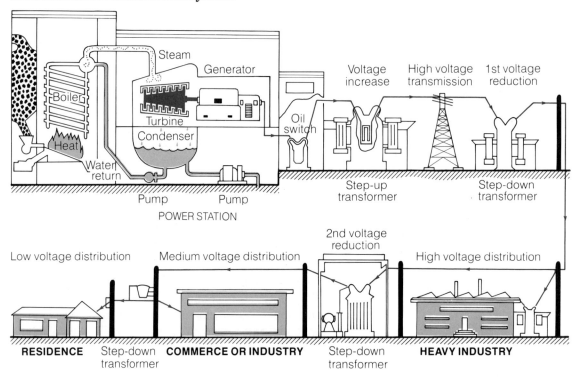

Electricity can be transmitted from one place to another very easily and rapidly. This is one of the advantages of electrical energy over other forms of energy. To better understand the transmission of electrical energy, the flow of electricity from the power plant to its use in a home or factory is traced on page 93.

The generators commonly in use today produce electricity at 13,800 volts. This is boosted as high as 700,000 volts at the power plant so that the electricity can be transmitted more efficiently. When electricity moves through wires, the wires tend to resist or try to hold back the flow of electricity. (This is called resistance.) It is important to keep this resistance as low as possible. High voltages (such as the 700,000 volts at the generating plant) have less resistance than lower voltages. This means that higher voltages can be transmitted more efficiently than the lower voltages.

Electricity is carried on wires to the area where it is needed. It is still at a high voltage. The voltage must now be reduced so the electricity can be used. This is done in transformer substations. Here transformers reduce the voltage to 4,000 - 25,000 volts. Some factories use electricity with voltages in this range. The voltage must be reduced again to 115/230 volts for use in homes.

Wires are an important part of the electrical transmission system. Wires usually are run overland on poles or towers. Some transmission lines run underground, but this rarely is done with high voltage lines.

This figure shows some of the more common types of poles and towers used in transmission lines. The single poles (a) are used

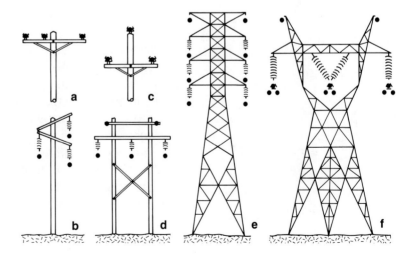

Energy and Power

How Transformers Work

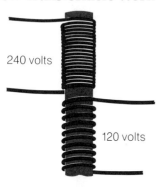

STEP-DOWN TRANSFORMER

EQUIVALENT TRANSFORMER

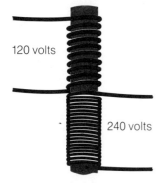

STEP-UP TRANSFORMER

for the lower voltage lines like those in residential areas. Other single poles (b and c) are used for midrange voltages between substations and often for the higher voltages used by industries. A wooden tower (d) is used to transmit higher voltages. Sometimes a wooden tower is used to carry more than one circuit. Metal towers are used to transmit the very high voltages from the power plant to substations (e and f).

Insulators are used on the poles to keep the wires from touching the poles or towers. Obviously the wires cannot be fastened directly to the poles. This is particularly true for the metal towers. Insulators are made of a ceramic material that does not conduct electricity. They are made so that they can be fastened to the pole or tower, and then the wire can be fastened to them in some way. On low voltage lines, the insulator is made so it can be screwed onto a wooden pin on the pole and the wire fastened to its side.

On high voltage lines, often there are several insulators fastened together. This is needed to keep the high voltage electricity from "jumping" the insulator and "shorting out" on the metal tower.

Transformers

As you have already seen, transformers are used to increase or decrease the voltage of the electricity that is being transmitted. These are called step-up or step-down transformers. Both work on the same principle.

Remember earlier in the discussion of generators, it was said that when a wire moves through a magnetic field, electricity is caused to flow through the wire? The electricity will also flow if the wire is held still and the magnetic field is moved. The effect is the same. There is movement between the wire and the magnetic field.

A transformer is made of two (or more) coils of wire that are not connected to each other. The coils may be wrapped around each other or just layed side by side. As the electricity moves through the wire in one coil, a magnetic field is set up. This is alternating current which is increasing and decreasing many times a second. This causes the effect of movement of the magnetic field.

The magnetic field from this coil is also surrounding the second coil. As the magnetic field moves around the second coil electricity appears to flow "through" the transformers although there is no electrical connection between the coils of the transformer.

When both coils of the transformer have the same number of

turns of wire, the voltage coming out of the transformer is the same as the voltage going in. But, if the second coil is smaller, the voltage will be reduced or stepped-down. This is called a step-down transformer. If the second coil is larger, the voltage will be increased, thus a step-up transformer.

Mechanical Power Transmission

Belt drive systems. Belt drives are made of a belt and two or more pulleys. One pulley is placed on some source of power such as an engine. The other pulley(s) is on an object being driven, such as a machine in a factory or the water pump on an automobile engine. The belt runs around the pulleys, transmitting the power from the engine to the objects being driven.

In most belt-pulley systems, tension is put on the belt. This causes it to pull tightly against the pulley. The friction between the belt and the pulleys keeps them from slipping. If the belt is loose, it will slip and the total power of the engine will not be transmitted. If the belt is too tight, it will wear out quickly.

V-Belt

Flat belt. The flat belt and pulley system is one of the oldest ways of transmitting power. The flat belt was used in early factories and textile mills to transmit the power from the same source of power to the machines. Many water-powered mills used the flat belt system. Later, after the development of the steam engine, long shafts were run through the factory. These shafts were connected to the steam engines by flat belts. The manufacturing machines were then connected to the shafts by more flat belts.

These early flat belts were made of leather. Though leather is used in some modern belts, rubber, cotton, and synthetic fibers such as nylon are also used. Often a fiber belt is coated with rubber so it will grip the pulley better and reduce slippage.

V-belt. V-belts are shaped just as the name says, in the shape of a *V*. The pulley has a V-shaped groove made into it. The belt is also shaped like a *V* and fits into the pulley groove. The friction between the belt and pulley is increased because the belt is forced in the V-groove. This allows the tension to be less on the belt without slippage occurring. Tension can be less than on flat belts. This also allows V-belt systems to run at higher speeds, with smaller pulleys, and less strain on the machines being connected by the system of belts and pulleys.

Cog Belt

Link belt. The link belt is a special belt made of many short sections. These short pieces are fastened together with metal rivets.

Link Belt

Exploring Technology

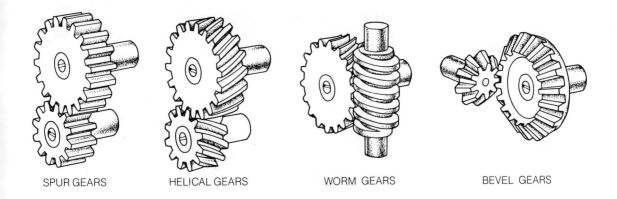

SPUR GEARS HELICAL GEARS WORM GEARS BEVEL GEARS

Gear systems. The first gears were made of wood and carved from round discs by hand. Modern gears are made from a variety of materials, depending upon their use. Iron and steel are used for strength. Brass can be used when the gear must be nonmagnetic, such as in a watch. Bronze, aluminum, and plastic are also used.

The spur gear is the simplest of the many types of gears. It is the least expensive and the simplest to care for. The teeth of spur gears are cut parallel to the center of the gear or shaft. Only one tooth is in mesh at any time. Thus, the strength of the spur gear depends on the strength of each tooth.

The helical gear is similar to a spur gear. Its teeth are cut at an angle (usually 45°) to the center line of the shaft. This makes the gear stronger. Usually more than one tooth is in contact at a time. This makes the gear stronger than only one tooth. The helical gear also runs more quietly.

Helical gears are made both left-handed and right-handed. The two are used together when power is being transmitted parallel to the line of the shaft. By combining two right- or left-handed gears, the shafts can be crossed at a 90° angle.

The bicycle provides one of the best-known examples of a chain and sprocket system. The bicycle chain is a roller chain. Notice the chain's sprocket. The sprocket is a toothed wheel similar to a gear. The links of the chain are made with openings for the teeth to fit into. This prevents slippage between the chain and sprocket thus reducing the loss of power.

Fluid power. Fluid-power systems transmit power through a substance such as oil, water, or air. In hydraulic systems, the fluid used is usually oil. In pneumatic systems, air is used. Both hydraulics and pneumatics are included in the term *fluid power.*

The basic parts of a fluid-power system are: a reservoir, pump or compressor, tubing or piping, cylinders, and control valves. The reservoir stores the fluid or air used in the system. It can also be used to filter dirt from the fluid or to cool the fluid. If needed, the reservoir can also be pressurized. A pump or compressor is used to move the fluid through the system and to provide the pressure when needed. Not all fluid-power systems require a pump.

Pipes, tubing, or hoses are used to carry the fluid to the various parts of the system. Lightweight tubing is used in aircraft where weight is important. When weight is not important, pipes are used. Flexible hose is used on moving parts or where there is much vibration. Over a period of time vibration can cause tubing or pipes to break.

Control valves have many uses. Some control the amount of fluid that can flow through the system. Others are used to control the direction of flow.

The cylinders are the power conversion units. They convert mechanical power to fluid power or the fluid power to mechanical power. To better understand fluid power, let's look at a commonly used fluid-power system—the automobile brake system.

Braking System of an Automobile

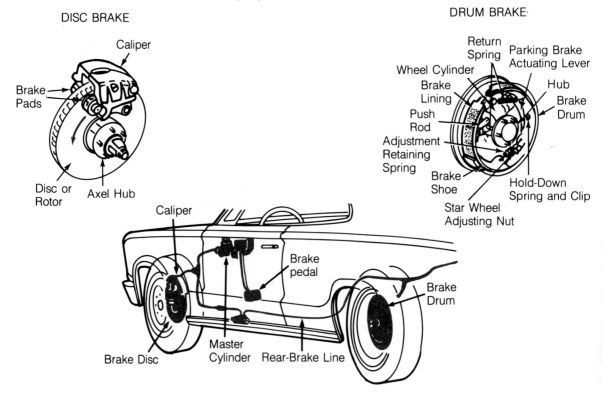

When the driver pushes on the brake pedal, he is pushing a hydraulic cylinder. In the cylinder, pressure is put on the fluid. This pressure is transmitted through the system of tubes and hoses. These connect to cylinders at each wheel. Thus, when the driver pushes on the brake pedal, he is applying pressure on cylinders at each wheel. These push the brake shoe against the wheels. The car then slows or stops.

Fluid power is used throughout technology. Did you know that a dentist's chair is raised and lowered by fluid power? Or that the power steering and automatic transmission on a car are fluid powered? The impact wrench used by a mechanic to remove wheel nuts when changing a car wheel is pneumatic powered. The heating and cooling systems in many modern buildings are pneumatic. Hoses or tubing connect the thermostats to furnace or air conditioner controls. Closers used to keep screen doors from slamming are examples of pneumatic cylinders.

Hydraulic power is used to lift the plow on a farm tractor. The landing gear of an airplane is raised and lowered by fluid power. The many machines used in construction—scrapers, rollers, bulldozers, and jackhammers are fluid powered. Fluid power is used in industries to control parts and assemblies as they move down the assembly line. Furniture industries use pneumatic sanders, drills, and screwdrivers.

Using Energy

It is important to understand the ways we use our energy. Much of the energy taken from our earth is not used efficiently. Much of it does no useful work. If energy does no useful work, it is considered to be lost or wasted. Of course, this loss cannot be stopped completely. But, it is possible to make better use of our limited energy resources.

In the first section of this chapter you learned that there are several sources of energy: fossil fuels, nuclear fuels, and the sun's direct energy, plus some other sources that are little used at this time. In this section you will learn some of the ways energy is used and how it can be used more efficiently.

About 26 percent of the total energy used in the United States is used to generate electricity. But, a majority of this energy is lost. Of the original amount of energy available in the fuels at the generating plant, less than one-half does useful work.

Some energy is lost in the first step of generating electricity. A certain amount of energy is used to change water to steam and to make this steam turn the turbines. The generator produces less electrical energy than the amount of energy that was put into the

Use of Energy in the U.S.

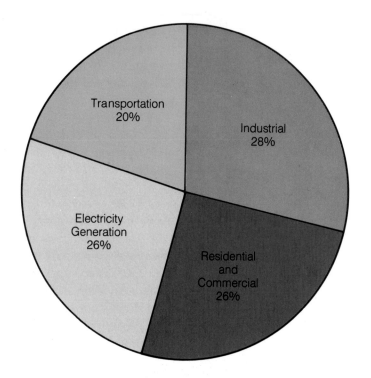

process of generating it. More energy is lost as the electricity is transmitted to the places where it is used. Still more is lost in the machines and the other users of the electricity.

The energy used in generating electricity accounts for only about one-fourth of the total amount of energy used. What of the other three-fourths? This is used in three major areas:
1. residential and commercial—lighting, heating, cooking, air conditioning, and running appliances in homes and businesses.
2. industrial—mining, processing, manufacturing materials into useful products.
3. transportation—fuel for the many different transportation vehicles in the air, on the land, and in the water.

Each of these major categories uses about one-fourth of the total energy used in the United States.

Some small amounts of energy are used in other ways. Some petroleum is used as raw material for plastics, fertilizers, drugs, and many other synthetic products. Though these uses are not major users of petroleum, they are important. Some, such as many important drugs and medicines, can be made only from petroleum.

It is important to understand that about half of the energy contained in original energy sources such as the fossil fuels does no useful work. Just as in the generation and transmission of electricity energy is lost before it can do useful work. Much of it is lost in the form of heat. Automobile and airplane engines produce tremendous amounts of heat. Heat is energy. This is energy that is not doing useful work. No machine is 100 percent efficient. No machine known today can make use of 100 percent of the energy put into it.

Heating—Most homes today are heated by furnaces fired by oil or gas. Some homes are heated by electricity. In some cases, a heat pump is used. This is a machine that operates much like a refrigerator. In the summer it removes heat from the house and pumps it to the outside. In the winter it can be reversed so that it puts heat into the house. Thus, this one machine is both a heater and an air conditioner.

Lighting in buildings is supplied almost entirely by electricity. Likewise, appliances are usually electric. Some water heaters, cooking stoves, and clothes dryers are gas heated but most household appliances are electric. Look around your home. Make a list of the different things that require energy to make them operate. How many of them require electricity? How many require gas?

Furnaces—Heating a building with an oil or gas fired furnace is less energy consuming than heating with an electric one. There are large losses of energy in the process of generating and transmitting electricity to the home. It is more efficient to take the oil to the home and burn it for heat. (The cost of transporting the oil to the home offsets this somewhat.) This is also true for hot water heaters and clothes dryers.

The all-electric home is an energy consuming home. The direct use of oil or gas for heating may use one-third as much energy as electric heat. Gas water heaters and clothes dryers may use one-half as much energy as electric ones.

Residential and Commercial Energy Use

Most of the energy used in residential and commercial operations is used to heat homes and business buildings. Modern construction methods cause us to use more energy for heating than older methods. The high-rise buildings are an example. It has been estimated that the Sears Roebuck & Co. building in Chicago uses more electrical energy per year than a city of 150,000 people. In buildings such as this, large windows or walls of glass lose heat in the winter. These large windows admit heat in the summer, making air conditioning necessary. Sealed buildings with windows that cannot be opened also make air conditioning necessary. Metal buildings transfer heat more rapidly than wooden ones.

Each year more and more electrical appliances are installed in our homes and businesses. There are more and more different gadgets that use more and more energy.

The amount of energy used by various appliances commonly found in the home is shown on page 99. Notice the difference between such items as black-and-white television and a color television; a frost-free refrigerator and one that is not frost-free. These are appliances that make our work easier.

Energy Use in Industry

Industry uses another large portion of the energy used in the United States.

natural gas	35 percent
coal	13 percent
petroleum	37 percent
electricity	14 percent

The metal processing industries are the largest industrial users. It requires great amounts of electricity to process bauxite ore into aluminum. (See Chapter 2, Processing) The steel-making industries require large quantities of coal.

Newer processing methods such as the basic oxygen furnace in steelmaking are more efficient. This process uses less energy to make the same amount of steel, but the older open hearth furnace can operate using up to 100 percent scrap metal. The basic oxygen furnace cannot. Therefore, the open hearth is still an important steel-making process.

The petroleum refining industry is the second largest industrial user of energy, accounting for about one-third of the industry total. It is difficult to measure exactly how much energy is used. The refineries use some of the raw material they are processing (oil and gas) as fuels for the process. Thus, some petroleum is used in petroleum processing. In addition a certain amount of the energy in the original petroleum is lost during the processing.

Many industries are working to reduce the amount of energy they use. They are improving processes in their plants to make them more efficient. They are taking greater care with the materials and energy they use. They are trying to eliminate waste. Many industries already have energy saving practices. The

ANNUAL ENERGY REQUIREMENTS OF ELECTRIC APPLIANCES

The estimated annual kilowatt-hour consumption of the electric appliances listed here are based on normal usage. When using these figures for projections, such factors as the size of the specific appliance, the geographical area of use and individual usage should be taken into consideration.

Appliance	average wattage	est. kwh consumed annually
Refrigerators/Freezers		
manual defrost, 10-15 cu. ft.	—	700
automatic defrost, 16-18 cu. ft.	—	1,795
Clothes Dryer	4,856	993
Washing Machine (automatic)	512	103
Washing Machine (non-automatic)	286	76
Water Heater	2,475	4,219
(quick-recovery)	4,474	4,811
Air Conditioner (room)	860	860*
Dehumidifier	257	377
Fan (circulating)	88	43
Fan (window)	200	170
Heater (portable)	1,322	176
Humidifier	177	163
Broiler	1,140	85
Coffee Maker	894	106
Dishwasher	1,201	363
Frying Pan	1,196	100
Hot Plate	1,200	90
Mixer	127	2
Oven, microwave (only)	1,450	190
Range with oven	12,200	1,175
with self-cleaning oven	12,200	1,205
Roaster	1,333	60
Toaster	1,146	39
Hair Dryer	381	14
Shaver	15	0.5
Sun Lamp	279	16
Radio	71	86
Radio/Record Player	109	109
Television, black & white, tube type	100	220
solid state	45	100
color television, tube type	240	528
solid state	145	320
Sewing Machine	75	11
Vacuum Cleaner	630	46
Waffle iron	1200	20

*Based on 1000 hours of operation per year. This figure will vary widely depending on area and specific size of unit.

Energy and Power

Hot pig iron is poured into an open hearth furnace. Metal processing requires the largest amounts of energy in industry.

reason for this is simple. The costs of energy are rising. Each energy saving technique saves them money.

As in other areas, the biggest waste of energy in industry occurs through loss of heat energy. Industries pump out huge amounts of heat into the air. Large amounts of heat are lost in water that is used in boilers or is used for cooling. This heat does no useful work.

If this heat could be used in some way, there would be less wasted energy. Some technologists have proposed that the hot water from steam turbine generating plants could be used to heat homes and businesses near the plant. This would get some work out of the heat energy in the water. Less energy would be lost. In addition, the fuel now being used to heat those buildings would be saved.

Energy Use in Transportation

One major use of energy in the United States is in the area of transportation. This does not count the energy used in producing the transporation vehicles. It takes into account only the energy

Use of Petroleum in the U.S.

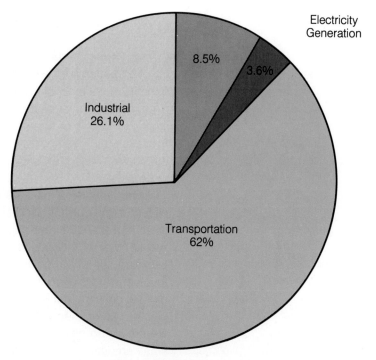

used by these vehicles, mostly the fuel they burn. The energy in the fuel alone is about one fifth of all energy used in the United States. Most of this is in highway transportation.

The transportation field is *energy intensive*. This refers to the amount of energy used for the amount of work produced. For example, trucks are more energy intensive than trains. Trucks use more energy to move one ton of freight one mile than the trains do. Trains are less energy intensive than automobiles. The autos use more energy to move each passenger one mile than do the trains.

Energy intensive is an important concept to understand. It is very important when we look at recent trends in transportation. Passenger cars use nearly two times the energy per passenger mile as trains. Yet in recent years in the United States, we have used cars more and trains less. We are getting less work out of the energy we are using, even though we are living in a time of increasing energy shortage.

Petroleum is used as the main source of energy for transportation. It is important to understand how we use it. Almost one-half of the oil we use is made into gasoline. This is used to run the engines in transportation vehicles.

Problems and Benefits

Conservation. When learning about energy and power technology, it is important to understand the problems as well as the benefits of the technology. One of these problems is conservation of resources. Most of the fuels used as sources of energy are limited. They are nonrenewable resources. (See Chapter 2, Processing.) This means there is a limited amount of these materials. When they are gone, there will be no more for people to use.

Some authorities estimate there is only a 30-40 year supply of oil left on the earth, if used at the present rate. It is estimated that there is a 200-300 year supply of coal left.

It is obvious that our fossil fuels must be conserved. We are using up our oil too quickly. As discussed earlier in this chapter, much of what is used is wasted. It either does no useful work, or is used to do work that is not essential. When we open a can with an electric can opener or brush our teeth with an electric tooth brush, we are using electricity to do work that is not essential. These small amounts of work can be done in other ways.

We are living in an energy-intensive world. This means that our way of living uses a lot of energy. Every day more gadgets that use energy are developed. Many are luxury items.

We have formed habits that waste energy. How many times have you turned on a light when it was not needed? How often are lights left on when no one is using them? Are light bulbs larger than necessary? During the oil shortage of 1973, many stores turned off half their lights. It was not as bright in the stores, but people found that they could shop nearly as well with the reduced light level. This saved energy.

During this same time, many people lowered the thermostats on their furnaces. They reduced the heat by only a couple of degrees. At the same time they used less heating fuel or electricity.

Tremendous amounts of oil are wasted in automobiles. Stand on a street corner during rush hour. How many cars do you see with one person in each? Just think, if each car had two people in it, there would be half as many cars on the road. Half as much gasoline would be used. This would be a tremendous saving of oil. If these same cars could get two times the number of miles per gallon of gasoline, the gasoline used could be cut in half again. If both these were done, one-fourth as much gasoline would be used!

Industry can conserve energy. Recycling of materials is an important way to conserve energy as well as materials. Some raw materials require large amounts of energy to change them into usable form. It takes more energy to make a ton of aluminum from bauxite ore than to convert aluminum scrap into a ton of new aluminum. This does not count the energy required to mine the ore or to transport it to the processing plant. It considers only the energy used in the processing.

Today steel mills are using an electric furnace that requires less energy than the type of furnace used only a few years ago. This is a tremendous saving of energy.

Benefits. In this chapter you have learned that energy and power are important to humanity. Our world is what it is today because energy and power were and are available for us to use. People over the ages have developed ways to use this energy and power. Work has become easier. The day's work has become less difficult. Less and less human muscle is required to do work. Many things are possible today that were not possible just a few years ago, all because energy sources other than human or animal muscles have been developed.

Look in a history book. What was the way of life 100 years ago? What was the form of transportation? How did people travel from one town to another? What types of homes did these people live in? How were these houses heated? How was food cooked? How did the farmers plow their fields, or harvest their crops, or haul them to market or store them?

All these tasks were affected by the sources of energy available to people living at that time. Compare this with modern times. How is all this work done now?

One can look at the importance of energy another way. Pretend that there is no electricity. What would you have to do without? What is in your home that could not be there if there were no electricity?

Consider the future. What would life be like if oil ran out tomorrow? What would you have to do without? Would there be

This potter is using a wheel powered by kicking a flywheel in the base of the potter's wheel.

automobiles, airplanes, trains, trucks? Are there any machines that can run without oil to lubricate them? How would you get to school, or go visit grandparents? How would your house be heated? How would electricity be generated?

Leisure time. Energy and power technology has made life easier in other ways. We have more leisure time. Work has been made easier. The work day and work week has also been shortened. More work can be done in less time because power is available to do it. Machines can do work faster than human beings or animals.

This gives you more time to do other things. In this free time many things can be done. Many hobbies are possible. Many of them are made possible by energy sources. What do you do as a hobby that is related to energy and power technology?

Model building is a popular hobby. Most models are of cars, trains, airplanes, trucks and ships. All are examples of machines made possible by power technology. Some of these use small power plants. These are made possible by power technology.

Many flying model airplanes are powered by gasoline engines. In model rocketry solid fuel propellant rocket engines are used.

Jet engines are made for model airplanes. Electric trains use electric motors to drive them. The switches, lights, signals, and even the whistle on the locomotive are operated by electricity.

Standard of living. Energy and power technology has helped to give us a better standard of living. In addition to easier work and more leisure, we also have better health. Our homes are heated better in the winter. They are cooler in the summer if we have air conditioning. We have luxuries never imagined 100 years or even 50 years ago. We live in a world of conveniences. We are entertained by radio and television. We listen to our favorite music group on a tape player or stereo record player. Cans can be opened with an electric can opener. One hundred years ago the can was not possible! You can brush your teeth with an electric tooth brush or comb your hair with an electric comb. You can polish your shoes with an electric polisher. You can play music on an electric guitar.

Suggestions for Further Study

Observations

Examples of energy and power are all around you. You can see them. You can learn by watching what goes on around you. The results and effects of energy and power technology can be seen especially on the environment. Make regular observations of a local electric generating plant. What color is the smoke coming from the smoke stacks? The darker the smoke, the more pollutants there are. What is the temperature of the water coming out of the plant? How does it compare with the temperature of the water up-stream from the plant?

Are there mining operations in your locality? What effect do these have on the area around you? Remember, there are both positive and negative effects. If the industry or mine were not in the area, there might not be work for the people there. Also, the industry or mine might be the cause of air or water pollution that is affecting your health. Look for both the positive and negative. Don't make your observations one-sided.

Make a survey of your home. What energy using devices can you find? How much energy does your home use in a week? Month? Year? Read your electric meter. How many kilowatt hours of electricity are used? See if you can find ways to conserve energy in your home.

Personal Involvement

You can learn by becoming involved with the world around you. The following suggestions will help you become involved in energy and power technology.

1. Read the local newspaper. Find how people in your area are conserving energy.
2. Build a solar collector. How can this device help conserve our supply of fossil fuels?
3. Make a survey of your home. How is energy used there? How can energy be conserved?
4. Find examples of energy waste in your community. Suggest ways this can be reduced or eliminated.
5. Work for a volunteer organization that is working to reduced energy waste.

Questions

1. How did change in sources of power affect the industrial Revolution?
2. What are the sources of energy?
3. How is the heat of fossil fuels converted to electrical energy?
4. How can water be a source of energy? Wind? Geothermal?
5. What do you predict will be the major source of energy 10 years from now? 20 years? 50 years? Will this change in the time period of 10 years to 50 years from now? Why?
6. How is the energy in petroleum converted to mechanical power?
7. What part of the energy in the various sources does useful work? How much is lost?
8. Name several ways energy can be conserved without making major changes in our life style.

Key Words

energy
power
solar energy
petroleum
fossil fuels
nuclear energy
undershot wheel
overshot wheel
hydroelectric
tidal energy
geothermal

turbofan
turboprop
rocket
transformer
insulator
v-belt
link belt
helical gear
roller chain
hydraulics
pneumatics

conversion
generator
turbine
internal combustion engine
turbojet
rotary engine

residential
commercial
transportation
energy intensive
hydronic
solar cell
photovoltaic

Testing Efficiency

Concept: *Efficiency*
When something is done with a minimum of waste, we say it is being done efficiently. When shopping, cleaning, or working, we can be efficient if we use our time wisely. Some machines are more efficient than others. They do work faster or need less energy.

Objective
In this activity you will build a simple machine and find its efficiency. You are trying to compare the amount of energy used by a machine with the amount of work it does. To keep it simple, energy will be shown in the form of weight.

Procedure: Construction
First, build the simple machine illustrated here. It is a shaft supported on each end. Drill a hole in each end for fastening the string. On one end of the shaft tie a string (24" long). Tie a washer to the free end of the string. Wrap the string around the shaft until the washer is touching the shaft. All the string should be wrapped neatly around the shaft. Tie the second piece of string on the other end of the shaft. A washer is attached to the free end of this string.

Procedure: Testing
You are now ready to test the efficiency of this machine. Let the washer on the string wrapped around the shaft fall. This will cause the shaft to turn. The other string will be wound around the shaft, lifting the other washer. The speed that the second washer is lifted is an indication of the efficiency of this machine.

1. Assemble the machine on a table for testing.
2. Tape a yardstick to the support nearest the end of the shaft with the test washer. The yardstick should be placed so the end of washer is at the zero mark.
3. Have a stop watch or watch with second indicator ready. You may want to have someone help you as a timer.
4. Count down and release the washer used for weight. Let it fall for ten seconds. Then stop the shaft from turning.
5. Measure the distance the test washer moved, if any.
6. Rewind the long string onto the shaft. Add a second washer to the end of this string for more weight.
7. Count down and release the weight, letting it fall for ten seconds. Stop the shaft from turning and measure the distance the test washer fell.
8. Continue this procedure, adding weight until you have been able to raise the test washer 12" in ten seconds.

Evaluation
Compare the number of washers needed to raise the test washer 12". Divide this number into 1. Example: If you needed 4 washers to raise the test washer 12", divide 4 into 1. 1/4 = .25. Convert this to percent by multiplying by 100. This test machine had an efficiency of 25%. The machine used four times as much energy as it created.

Does this surprise you? Automobile engines are only about 37% efficient. Could you improve on the efficiency of the test machine? How? Try to make a test machine similar to the one above that is more efficient.

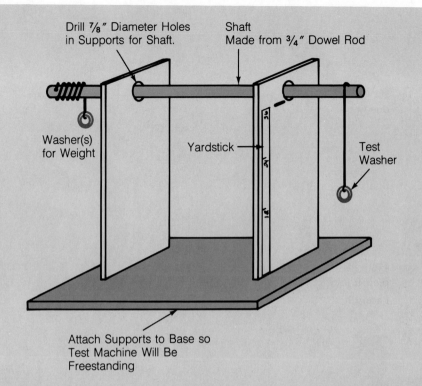

Energy and Power

109

An employee monitors the production of synthetic fibers.

Chapter 4   Manufacturing

Introduction

Once raw materials are made into usable form (processing), they are made into a product. Sheets of steel are made into wastebaskets, auto bodies, and fenders for bicycles. Lumber is made into chairs, tables, and toys. These are things we see and use every day. We tend to accept them and take them for granted. But these items have gone through a complex manufacturing process. Each was carefully researched, designed, planned, produced, and marketed.

Modern manufacturing began during the Industrial Revolution from 1750 to 1850. About this time interchangeable parts were first used. Eli Whitney is credited with first using interchangeable parts in making guns. Each part was made accurately. All triggers, for example, were made alike. Before this time each gun was made separately by a single craftsman. Each gun was different from all others. If a part in a gun broke, a new one had to be made to fit that gun. By using interchangeable parts, a broken part could be replaced with an identical part. Spare parts could be made and stored until needed.

A new way of orgainzing the worker was also developed during the Industrial Revolution. The assembly line helped speed production. Each worker was given one small job. The object being made, moved along from one worker to the next. As the object passed each worker, another job was done on it. A part might be added, bolts tightened, or the item tested. This specialization of labor is important in manufacturing. Each worker is a

specialist. Each does only one small operation. So, modern manufacturing depends on three important principles:
1. Interchangeable parts
2. Assembly-line method
3. Specialization of labor

Modern manufacturing is a very complex process. There are many steps in manufacturing a product. Many resources are needed: machines, raw materials, buildings, people to do the work, and money to pay for these.

An idea must be designed and developed into a product that can be sold. The product must then be produced. Once made, the product must be marketed. Advertising is used to tell people about the product. Then a system must be set up to get the product to customers. Each of these steps will be studied in this chapter. After studying this chapter, one should have a better understanding of manufacturing systems.

Resources of Manufacturing

Production resources are materials, equipment, knowledge, human power, energy, and money or finance. These resources are needed during the manufacturing process. Production of a product requires different amounts of each resource. For example, the production of tires requires a variety of materials, equipment, and knowledge.

Each resource is dependent upon the other for total production. This means that each resource must be available before the product can be produced. Without money for financing, material flow and labor would stop. Without equipment, the materials could not be processed into the finished product. Therefore, we can say there is a relationship among resources in production requiring all resources to be present before the product can be produced.

Finances

A company that produces products must get money or financing for such items as equipment and materials. The financing for this production can come from several places. A company can borrow the money from a bank and pay interest on it. Interest is money paid to the lender for borrowing the money. It is an expense that must be added to the total cost of production.

Other sources of financing are Small Business Administration loans or the sale of shares of stock in the company. Selling shares or parts of the company does not mean the company is selling machines or buildings. It means the company is selling a part of

HISTORICAL DEVELOPMENT OF MANUFACTURING

Time	Development
3500 B.C.	Use of wheel and axle for transportation
500 B.C.	Lathe used for woodturning
1569	Screw-cutting lathe developed—Jacques Besson
1769	James Watt invented steam engine—later used to provide power to industry
1774	Precision cylinder-boring mill developed—John Wilkinson
1790	Samuel Slater opens first successful textile mill in the United States

continued

HISTORICAL DEVELOPMENT OF MANUFACTURING

Time	Development
1793	Eli Whitney builds first cotton gin
1798	Eli Whitney invents milling machine to produce standardized parts of muskets
1801	J.M. Jacquard invented silk loom—punched cards control the machine
1851	Isaac Singer patented his sewing machine
1900	High-speed steel cutting tools developed
1903	Oxyacetylene welding torch developed
1903	First fully automated machine-made bottles
1907	Paint spray gun developed
1913	Ford Motor Co. opens first moving assembly line

continued

the business. This makes the shareholder a partner with the company. This partner is also called a *stockholder* and the company that sells shares of stock is called a *corporation*. Sears Roebuck & Co., General Motors and RCA are examples of large corporations. Their shares are bought, sold, and traded on the stock exchange every day.

The value of these shares will change each day. One possibly could buy a share of RCA stock at $38.70 today and its value could reach $40.50 by the end of the same day. Should one choose to sell at the new price, one would have made a profit.

Stockholders expect to make a profit when they buy corporation shares. Corporations sell more shares to get more financing. Thus the corporation gets more money to expand. Stockholders get *dividends* (money) from profits made by the company. The stockholder's shares of stock may increase in value as the company prospers or decline in value if it doesn't.

Human Resources

Human resources of production refer to the people who work to make the product. These workers are called *personnel*. The personnel are divided into two divisions, *labor* and *management*.

Labor jobs are the production ones, for example, machine operator, sander, cut-off-person, or painter. The *wage* or payment is by the hour and upon experience, amount of time in that job, and the value of the job. Some jobs are very highly skilled,

Jobs in manufacturing require varying amounts of skill. *left* A stitcher in the garment industry. *right* An electronics technician.

Manufacturing

HISTORICAL DEVELOPMENT OF MANUFACTURING

Time	Development
1914	Centrifugal casting of cast iron pipe—re-usable molds are used
1920	Ford introduces continuous casting of cast iron for engine blocks
1921	Jigs & fixtures used in the jig-boring machine to make rifles and revolvers—Enfield, England
1923	First wide strip mill for making sheet steel
1930	First automatic factory—Made chassis frames for cars: one every six seconds
1942	Liberty ships—cargo ships built in 10 days and ready to go to sea five days later
1955	Artificial diamond for use in cutting tools is developed
1962	First industrial robot
1963	Electro-coating methods for painting car bodies is developed
1964	Technique for fast-braking electric motors developed—machine tools can now be stopped quickly
1985	First products manufactured in space went on sale—tiny plastic beads, perfectly round and uniform in size

such as an electronic technician or machinist. Other jobs require little or no previous experience, such as stock or shipping workers. These workers would be trained to stamp out-going packages or to bring stock items to the assembly line.

Both types of workers are necessary for production. But the skilled job requires years of training, whereas a high school graduate easily could get employment as a stock or shipping worker. The hourly wage for the skilled employee could be from $10.00 to $20.00, whereas the stock or shipping worker would start working at about $5.00 per hour.

A *piece rate* or the number of items to be made per hour is set for normal production. For example, a sewing machine operator must install sixty zippers to meet production set by the manufacturer. The worker who exceeds the sixty would be paid extra for each zipper installed. Some production workers are paid by the hour. They usually stock the assembly line or inspect the products.

Management employees are responsible for the running of the company. They perform such work as keeping pay records, ordering materials, distributing the product, and making decisions about what product to produce. (This side of personnel is usually called top management or the boss.) The management positions are placed on a chart called an *organizational chart*. This chart tells who is responsible for each part of the company or to whom to go to see about designing, engineering, distribution, or personnel.

Management employees are not paid by the hour but by the month. This type of payment is called a *salary*. Management employees are not paid for working overtime like laborers. They are paid a set salary no matter how many hours they work. This means they receive a very stable income.

The job titles under management, for example, could include accountants, designers, engineers, or president. These workers are usually college graduates who study materials and processes so they can help produce better products. They usually are trained in business and engineering. Often a group of workers will work years to produce a new product. This type of work is called *research and development*. Research and development has given us many new and improved products such as color television in place of black and white, ten speed bicycles in place of the standard one speed, or electronic calculators in place of the adding machine.

Organization Chart for a Company

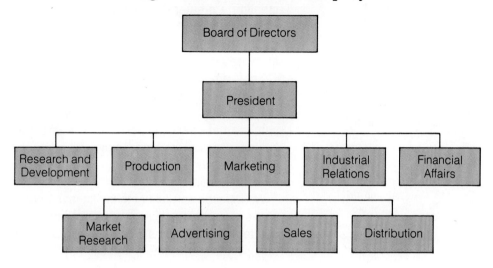

Natural Resources

Resources are the materials used to make the product. Steel, aluminum, wood, glass, coal, and paper are examples of resources. See Chapter 2 for more information on the processing of raw materials. Energy to change these raw materials into finished products is also a valuable resource. Coal is burned to produce steam. Steam is used to generate electricity. Natural gas is the energy source in making glass. In manufacturing, materials are taken into a factory and made into a useful product.

Resources are important in another way. Quite often manufacturing has a bad effect on our environment. Often the earth is spoiled when materials are taken from it. For example, when coal is mined in strip mines for fuel, the land is left scarred. Some factories pump polluted water into lakes and rivers or put fumes and smoke into the air.

This is a serious problem for manufacturers. Materials are needed or industry cannot operate. These materials must be obtained as cheaply as possible. To clean up the land after strip mining is expensive. To remove the impurities from the waste water is also expensive. The clean-up operations add to the cost of the material.

For many years most manufacturers were small. There were not many industries. The amount of waste products from industries going into the environment was small. But populations grew, creating a greater demand for products of industry. As a result, industries grew. More and more raw materials were needed to supply these industries. Therefore, more and more waste materials were put into the environment.

Small amounts of waste can be absorbed by the earth's natural systems. Put one drop of food coloring in a quart of water. You hardly notice the change in color. But, put ten drops in and you can see the effect dramatically. The effect of small amounts of industrial pollutants was small. But when many industries put large amounts of waste materials in the environment, the effect became great.

We must be careful not to think that manufacturers are creating all the environmental problems. This is not so. We are learning that our technical systems are very complex. Energy, power, and transportation systems, production systems, and communication systems depend on each other. They are interdependent. This means each depends upon the others. Materials are processed and transported to the manufacturers. Communication is necessary to keep all the systems running smoothly. Manufactured products must be transported where they are needed.

Energy Resources

Energy is a resource needed in manufacturing. Modern industry must have energy to run machines. Buildings must be heated. Energy is needed to convert raw materials to a usable form. The glass industry uses large amounts of natural gas. Steel mills need heat to melt the iron ore to make the steel.

As production increases and industries grow, each year the need for energy increases. Increases in energy needs are placing a drain on the world's oil and gas deposits. How much oil and gas is there in the United States? Will we run out of fuel? What will happen to future generations? The answers to these questions are of major concern to manufacturers. Conservation of energy resources is important in manufacturing.

Atomic energy is a possible answer to energy needs in manufacturing. But it is not without its costs. The building of an atomic-powered generating plant requires an isolated location and elaborate safety precautions. The biggest problem to solve is what to do with the radioactive waste material from the plant. The waste material has to be buried deep in the ground in special containers. Even this solution is presenting problems.

Scale models are made before products are manufactured. By making the model as accurate as possible, designers can see what the product will look like. Design changes can be made before the production line is set up.

Product Development

In manufacturing, ideas for products come from research departments, product designers, and engineers. The researchers try many combinations of ideas, materials, and techniques for producing a product. These are used by the designers who make sketches, drawings and models or work-ups of possible products. Their ideas first begin as sketches and drawings. Models of different designs are produced.

The product engineers provide technical information about the product and the ways to produce it. Often prototypes of possible products are made to test the usefulness and working of a product.

Once all the needed resources are gathered, the product can be developed. Millions of dollars may be spent developing a new product. This is carefully planned. The steps of product development are usually: (1) design, (2) engineering, (3) market analysis, (4) production planning.

Design

An idea for a new product is difficult to develop. Some of the newest product ideas come from toy manufacturers, who must develop new products to continue growth. Other manufacturers develop new products also. For example, Polaroid Corporation developed a new camera that gives an instant picture. Is this a new product? The first box camera was invented in 1888 by George Eastman long before the first Polaroid. However, the instant film used by Polaroid is a new product. It was introduced and patented by Edwin Land in 1947.

Can you think of any new products? Many new products are improvements on old ideas. Automobiles, coffeemakers, hairdryers, cameras, and calculators are examples of products being improved regularly.

Designers create new product ideas and change old products. Many designers use the computer in their work. Computer assisted design (CAD) programs allow the product to be designed on the computer screen and stored in computer memory for future changes. Designers can look at the design on the computer screen from different positions. It can even be viewed in three dimension. When all changes have been made, final drawings of the object can be made on a plotter. This process eliminates the time normally needed for drawings of the object as it is being designed.

As the staff of designers develop the product, different teams work on different parts. In designing an automobile, for example, different teams work on areas such as the body, the engine, the steering, and the interior. Drawings for each part can be viewed on the computer screen, so each team can see what other teams are doing. Management can get a view of the total project as it develops. All this can be done while sitting in front of a computer screen.

This CAD systems allows the designer to view a compact car in three dimensions.

The term CAD is also used to mean computer assisted drafting. This is closely tied to the computer assisted design discussed above. As designers work to design a product, their ideas must be made into production drawings. Drafters make drawings of the individual parts of the product. These drawings are used to tell the production workers how to make the parts and assemble them into the finished product.

Product Research
Much research goes into the design of a new product. It begins with someone's idea. Then the idea is refined through numerous sketches and drawings. Researchers must determine the following: What materials can be used to make the object? Which materials are best? How can the product be made? What machines can be used? How can these machines be arranged most efficiently? All these questions must be answered.

The cost of the product must be kept as low as possible. So, the best material may be the one that is the least expensive. Also, the cost of labor must be kept as low as possible.

Research into a product's past performance or the development of a new material may cause a product to be redesigned. For example, how many products can you think of that once were made of metal but are now being made of plastic? The tricycle is a product that has many parts now made from plastic. Many

Researchers monitor an experiment in the laboratory.

Manufacturing 119

household appliances like mixers and vacuum cleaners are now plastic.

Plastic is a designer's delight. It can be shaped into more different shapes than other materials. Plastic is also available in many bright colors. Many materials used in products are being replaced by plastic. Why? Cheaper production is not always the reason. Some products can be reduced in weight, can be used longer, and have improved performance by having plastic parts.

Research on products is conducted to help make improvements in them. Tire companies constantly are testing and retesting their products to improve mileage and safety.

New designs of old products usually result from research and development. Companies who make the same product are in competition with each other. This competition may cause each company to redesign their product each time their competition introduces a new model.

Television manufacturers are always competing for customers. Each year a company may make changes that forces the other manufacturers to do the same. For example, when the first color set was introduced, television manufacturers had to develop their own color model. Some companies became leaders in their field as a result of this competition. Such companies have a very good research and development team that give them the advantage.

Using this computer-aided workstation, an engineer designs electronic circuit chips. Using the computer, the design takes less than six months. Without the computer, it took 18 months or more.

Market Survey

What company can introduce a product on a guess that the public will buy the product? No company can. Before the new product is introduced, a prototype of the product is made. A *prototype* is the real product, but only a few are made for testing and market analysis.

Market surveys are conducted in average communities. People are asked if they like the product. The survey is taken among families who might buy the product. From the results a company can decide, first, if there is a market for their product; and, second, how big the market may be. The company now can decide to produce and how many products they should make.

PRODUCT SURVEY

The answers to the following questions will be used in making a product. Your help in filling out this survey is greatly appreciated.

Survey No. _____

1. Age_____

2. Male_____ Female_____

3. Do you own a similar product? Yes_____ No_____

4. How does this product compare with yours, if you now own one?
   Better_____, As-Good-AS_____, Not as Good_____

5. Would you consider purchasing this product? Yes_____, No_____

6. If yes, what would you pay?
   $1.00 - $2.00_____ $2.00 - $3.00_____, $3.00 - $4.00_____
   $4.00 - $5.00_____, $5.00 - An Over_____

7. Would you consider buying more than one if there was discount?
   Yes_____, No_____

8. How would you plan to use the product?
   Gift_____, Personal Use_____, Family_____, Friends_____

9. If you don't plan to purchase now, please check reasons.
   Need to Save Money_____ Too Expensive_____
   Don't Like Design_____ Don't Need One_____

10. Under what conditions would you consider purchasing?
    Lower Price_____ Better Quality_____ Home Trial_____
    Time Payment Plan_____, Money Back Guarantee_____

Feasibility Study

Once the product has been designed, tested, and the market study made, a final study of the company's ability to produce the product, the *feasibility study,* will help management make the decision to produce. The feasibility study consists of the following:
1. The history of the market for the product.
2. The company's present operating capacity.
3. The anticipated demand for new equipment, materials, and labor.
4. The future of the product.

The results of this study take a great deal of the guess work out of producing a new product. Management can say with some certainty that the new product will make a profit for the company and its stockholders.

Producing a Product

Production is the process of making the product. There are many processes involved in the production of a product. The automobile has many different materials that have to be formed and finally assembled into a car or truck.

There are several steps in producing a product:
1. In *plant layout,* the factory must be designed and machinery organized. The assembly line must be arranged.
2. In *materials handling*, the materials used in making the product must be transported into the plant. Parts must go the correct

In this automated body assembly area, 18 pairs of robots apply spot welds. 93% of the total 5,000 welds made are applied automatically.

places on the assembly line. The finished product must be taken out of the plant.
3. Many of the materials are made into parts. Sheets of metal are made into doors, fenders, or trunk lids. This is *materials working*.
4. During all phases of production, the work is inspected. This is *quality control*. Inspectors check to see that the parts are the correct size. They inspect welds, paint, and every step in the assembly.
5. Assembly is the most dramatic part of manufacturing. At the beginning, one part is put on the conveyor. As it moves, parts are added. In auto assembly, the frame starts first. The engine is added. Body, fenders, doors, seats, wheels are added. At the end, gasoline is added, and the car is driven out.

Jobs in Production
One way to learn about the total production operation is to study the work people do. There are many different production jobs. A few are described here. Reading about these jobs should help you understand more about manufacturing industries.

Manufacturing engineers are responsible for plant layout. This includes every phase from the delivery of raw materials to the storage of the final product. Everything must be considered—utilities, equipment needed, handling of materials. To perform this big job, the engineer uses a floor plan of the plant where the product is going to be produced. Then each piece of equipment is given a place on the drawing.

The best location of equipment is decided by what it does and when that part of the production needs to be done. Furniture manufacturers are very careful about keeping the machines away from the finishing or the upholstering area. Once the equipment has been put in the best location, the next job is to arrange for the movement of material. There must be a steady flow of material to each machine or work station. To determine this operation, a material flowchart is developed. The flowchart can be simple or complex, depending on the number of operations and part of the product. A simple product like a wooden pencil could require a large number of operations. Some of the information included on the flowchart is: (1) the operation to be done and its number, (2) where the product goes next, (3) if it is to be inspected, and (4) where the item is to be stored. (Symbols are used to represent each of the operations.)

Flow-charting a product is very interesting work. Figuring out the where, how, and when of a product can be quite a challenge. Each part of the product will be routed to an assembly. A good example of this is the automobile assembly line. The number of individual parts in an automobile is about four thousand. All parts of the car have been flowcharted at one time or another. Some parts of the automobile are assembled in a separate plant and shipped to the assembly plant. These are called subassemblies. An automatic transmission is a good example of a subassembly. All the parts are supplied in the quantities needed to make the transmission in another plant. The production rate for midsized cars at General Motors is fifty-five cars per hour. When you think about making or assembling fifty-five cars in one hour, it does not sound possible. This production rate can only be accomplished with a massive conveyor belt over five miles long.

Each part of the auto being made is placed on the conveyor at some time along the conveyor. This makes it possible to custom make cars to order. An order for a car with red interior, bucket seats, AM-FM radio, white top, red bottom, and two-door model would have its special parts set on the line to be assembled as

Flow Charts of Parts and Materials

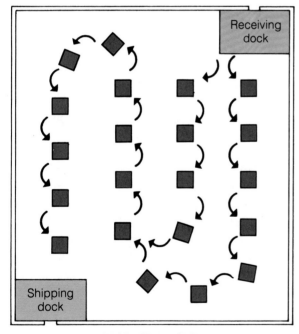

PRODUCT LAYOUT

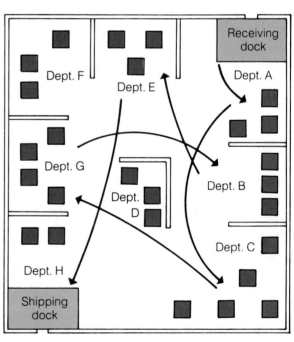

PROCESS LAYOUT

that car goes by the work station. Automobile assembly is one of the most automated lines in the world. Computers are required to keep things organized—to help in planning that the right parts are at the right place when needed. Yet, it still requires thousands of employees to work on the five-mile line. Job tickets, listing the parts needed, are attached to the car. Workers can check these tickets to be sure they are putting in the correct parts.

Just-in-time assembly is now being used in some assembly plants. "Just-in-time" means that a part arrives at the assembly line just as it is needed. Though this requires a lot of organization, it results in cost savings. Few parts are kept in storage. They are scheduled to be shipped from suppliers to arrive when needed. The scheduling is done by computers. When the system is in operation, very little warehouse space is needed at the factory. This reduces the cost of the product.

Tool and Die Maker
Making special tools for production is the job of the *tool and die maker*. A production tool could be a tool designed to help production. An example could be a special wrench to be used on the assembly of a car or appliance. These special tools are also required to repair or service the product.

A die made by the tool and die maker usually requires a detailed drawing of the part to be made. A die is a permanent mold, which means it can be used over and over again to make the same part. An example of a die would be the mold used to make frisbees. This die is used over and over again to make each one the same. Dies are interchangeable in machines so other parts could be made on the same machine. Often a producer will fill an order for 10,000 units of a single part, then change dies to fill another customer's order.

Many tool and die makers are highly trained machinists who then made a change into tool and die making. Other jobs of the tool and die maker are to make jigs and fixtures. These items are used to hold the parts while machine operations are performed. An auto engine block or a bicycle frame must have all holes drilled in the same spot on each model. This is necessary for interchangeability of parts to assure proper fitting.

Machine Operators
Operating the machines of industry is the single biggest job in manufacturing. The positions cover a wide range of machines. Some workers use machines that are very simple, such as a lug

nut installer. Others are highly skilled technicians who inspect or test electronic components, using complex machines.

Machine operators are under a great deal of stress because they must operate their machine to capacity to meet production goals. A time and motion study is conducted to see how much time an operator takes to make a unit of the product. Time studies are done by the industrial engineer. All machines on production jobs must be time-studied to determine total production time and how to pay employees based on their production rate. The products have to be inspected by inspectors to make sure all parts and all final products are the same. On high quality jobs, this may be done as often as every hour. One might see this being done on a luxury automobile. Any adjustment to be made requires the machine to shut down, which cuts production and reduces the worker's pay.

When CAD (Computer Aided Design) is used to design parts, it is possible to connect the computer into the manufacturing stage of production. This is referred to as CAD/CAM - Computer Aided Design/Computer Aided Manufacturing. The computer program used to design an object can also be used to control a machine (or machines) needed to make the object. A milling machine or lathe can be controlled by the computer. The part can be made very accurately, and human error is reduced.

As automation of this type increases, fewer machine operators are needed. Many of these workers are retrained to operate the automated machines being used in manufacturing.

Computer Programmer
Computer programmers are becoming an important, highly skilled person in manufacturing. They develop the programs needed to operate the computers being integrated into many manufacturing plants. As more computers and computer-controlled machines are used, there will be increased need for workers trained to use the computers.

Assembly Workers
Robots are being used more each year on assembly lines worldwide. Auto makers are redesigning their production lines to include robots and other computer controlled machines. The robots can be programmed to pick parts from a conveyor and place them where they are needed. Bolts and nuts can be put in place and tightened. Welding can be done automatically and very accurately. Jobs that are dangerous for humans can be done by robots. In

All of the wheel lugs are fastened in one operation with a special tool.

a spray booth, for example, fumes can be very toxic to humans. Robots can be programmed to do the spraying, eliminating the hazard to humans.

Robots are used in the electronics industry to assemble circuit boards faster and more accurately than humans. Increased speed and accuracy result in cost savings. A better product is produced at a lower cost.

See Chapter 8, "Automation," and Chapter 9, "Computers," for more detailed information on these two topics.

Spot welding on the body of a car while it moves down the assembly line.

Manufacturing

Quality-Control Inspector

Quality control is not a production job. The quality-control department makes checks on all phases or parts of the production. Parts are checked to see if they meet standards set during the set-up of the production. Checking parts to make sure they fit properly is only one step. Materials are also tested to assure proper strength. Sometimes as the materials in use change, parts weaken and a breakage will occur.

Quality control at an automobile plant means every car is tested for proper braking, steering, lights, and equipment operation. Every tenth car is pulled off to the side and inspected very carefully for any flaws in assembly or materials. For example, should several problems occur in the upholstering department then the foreman is called in to see where the crew is having trouble. The problem may be solved easily once it has been identified. A defect in materials could cause a safety hazard. Sometimes auto manufacturers have to recall many cars to correct defects discovered after the cars have been distributed.

The testing of materials is conducted in laboratories under scientific conditions. The results are compared with the standard set. In some cases a strength test is performed to see if the bumper metal meets standards. Another example is testing paper for strength as it comes off the machine where it is made. If the paper meets standards, then the paper machine goes faster, and miles of paper are produced.

Computer display terminals must be inspected and tested before shipping.

Exploring Technology

Quality of the product also determines, to a great degree, the success of the company. The results of unhappy customers can destroy any company. Management works to create a positive attitude or pride in the work among employees. The zero defect program is an example. Zero defects means all final, assembled products will be good products. The zero defect program was developed by space contractors who were making parts for the space program.

Any defects in equipment or materials in space could cost the entire mission. Employees working in zero defect are being asked to make sure all products meet certain standards. This does not mean that there are no defects.

Safety

Safety is of concern to all employers and employees. The concern of the employer is to protect the employees from injury. Injury on the job costs the workers time from the job in addition to the pain of the injury. Nobody wants accidents to occur. An accident could mean the loss of life, limb, or eyesight.

Responsibility for accidents has to be shared by the worker and the employer. The working conditions for the employee are the responsibility of the employer. The use of safety devices, following safety rules, and using common safety sense are the responsibility of the worker. Employers develop safety programs for their employees. The person in charge of the program is the safety engineer. This job requires the inspection of the following:

1. All machines for safety
2. Buildings for fire and ventilation
3. Materials used in production
4. Safety equipment used by employees

National regulations that govern companies and employees were written into a law, the Occupational Safety and Health Act. This law requires companies and employees to meet certain minimum safety standards. Companies not able to correct safety hazards have to stop operations. Employees have to obey new rules or be fined or lose their job. The overall effect of the law has been to reduce accidents and deaths. The price for this reduction was high, but the results mean less time lost because of injury and a safer environment for workers.

Safety programs for employees are a part of the management's responsibility. The program is designed to create a positive attitude about safety. This is accomplished by stressing the positive side of safety and giving safety awards. Uniform safety

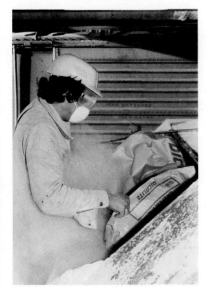

A safety mask and goggles are necessary equipment when working in areas of fine dust or fiber particles.

standards for miners, automobile manufacturers, truck drivers, and farmers are a big job. Making the mines, factories, roads, and fields safe for the worker is the goal.

Marketing

Mass production requires mass sales. An industry can make thousands of a product. But if no one will buy it, the industry will not stay in business very long.

Marketing is more than just selling. Market surveys are made. A system for getting the product to the buyer is needed. This is called *distribution*. Advertising tells the buyer about the product. Finally, the product is sold.

Market Survey

Market surveys have been discussed as they relate to the planning of a new product for production. Market surveys are also conducted to determine when to introduce a product, who would purchase it, how much the product is worth, and the need for the new product. This material about the market is used to help sell the product. Planning when to introduce a new product can be very important. Is the consumer ready for an electric toothbrush or electric car? Some markets are seasonal, for instance, a new golf club would be marketed in the spring. A key time for maketing toy products is Christmas.

Computers have been an important tool in many business applications including marketing, accounting, sales and advertising.

Exploring Technology

The people who are most likely to buy a product are identified through surveys of the community. Many questions about the need for a new product are answered in a survey of this type. Questions like the following are asked:

1. When would you buy this product?
2. Where would you go to buy the product?
3. How much would you spend, ($10.00-$20.00. $20.00-$30.00)?

Distribution and Sales
The delivery of a product to the consumer is called distribution. Products can take many routes to the customer. Let's examine a few ways you can buy a product.

1. Buying from a department store. (Retail)
2. Buying from a company that normally sells to businesses. (Wholesale)
3. Buying from the factory that produced the product. (Direct)

Retail sales for a large corporation like Sears and Roebuck would total $11 million a year. Sears sells in large volume which means that large purchases of the product from the manufacturer are necessary. To sell in large volume, a company needs a large storage area and a large sales staff to sell many products. The price of each product must include the cost for storage and staff cost. This expense is called *overhead*.

The distribution of products like televisions, auto parts and building materials is done both ways, retail and wholesale. Wholesale distribution means the customer is buying the product for resale to someone else.

A television manufacturer sells the sets to an area distributor, who in turn sells them to retailers. The wholesaler is providing a service to the smaller retailer that the factory cannot provide. This service is selling small amounts and delivering products on demand of sales.

Wholesalers normally sell only to retailers or other businesses. There are several reasons for this restriction from selling to the public. Wholesalers are organized to handle large purchases and deliveries, whereas an individual may buy only one product. Often the retailer places the order by phone to the wholesalers and there is no showroom for the public to view the product.

Some companies sell to both the public and large businesses. The public pays one price and the businesses pay a reduced price because they buy large quantities.

Factory Outlets

Manufacturers of some products like shoes, clothing and glassware often sell directly to the public at prices between wholesale and factory cost. These factory outlets are usually located in or near the factory. For this reason, only a few people can take advantage of the savings. This service to the public is usually a way of selling *seconds* or less than perfect products.

Distribution of products to the wholesalers and retailers is the final link in the manufacturing process. The products are transported by trucks, trains, planes, and ships. Cost of delivery is added to the cost of the products.

Advertisement

Advertisement is a form of communication. The message is that the product is the best for you. Once you are convinced that you need the product, the price becomes secondary. Modern advertising may be done through radio, television, newspapers, magazines, signs, and circulars.

Sales promotions or a drive to sell a large quantity of the product begins with an advertisement. The advertisement about the sale or promotion is designed to highlight the very best features of the product. The key to the promotion is to advertise the sale in several media to get a large cross-section of the population. A tradition that has developed across the United States is to have sales on special holidays such as Washington's birthday, Fourth of July, or any holiday when people do not work. Other countries have similar sales on their national holidays.

Advertisements on television are very expensive but reach a very large audience of potential buyers. This form of advertisement usually is directed toward creating knowledge about the product so the buyer will choose it instead of another brand. Advertisements on television are highly developed to tell a message in fifteen to sixty seconds. The message is on the level of the consumers and is played during the time when the right age group will be viewing. Television ads cost according to the time of day and the show on which they appear. Prime time (evening hours), coast to coast, television time may cost from $125,000.00 per minute. Other types of advertisments may be on pencils, pens, and calendars; they serve the purpose of telling the consumer about the product.

Shoppers are persuaded sometimes to buy one product over another because of the packaging of the product. Packaging the product in the most attractive way is a form of advertisement. This

Manufacturers use various promotional techniques to attract customers to their products.

is particularly true of food items that have to be sealed for freshness. Can you remember shopping for a new brand of candy bar? What you purchase may be the best-looking packaging.

Advertising of products is a big business. Other functions include:
1. Changing the company's name, for example: Esso to Exxon.
2. Changing a corporation's image by telling the public how their products help everyone.
3. Establishing a new brand name for the public to identify and seek when shopping. Brand names like GE, RCA, and Zenith televisions are more likely to be purchased than less-known brand names.

Product sales can act like highway and speed limits. When sales are high, the speed of production is high. Likewise, when sales drop off, the production rate must drop or change to another product. This principle of economics is called the *law of supply and demand*. The aim is to have the supply of the product meet the demand of the product.

Some products would go down in price if the company produced too many items. Such items as food, clothing, and cars, are subject to this decreasing value. Food is also subject to increase in price if the demand for the product increases or the quantity of food is decreased.

Sales, as have been seen, are controlled by the supply and demand for the product. There are other factors that affect sales like the time of the year or season, price, and location. Time of the year controls to a great degree the sales of items like suntan lotion, cold medicine, lawn mowers, or sporting equipment. Christmas is a season when lots of new products are introduced. Seasonal sales require stores to change continually the product displays. This involves trained decorators to design and arrange attractive displays.

Price
Pricing of an item for resale is a very competititve business. Companies have personnel who shop with their competitor to compare prices. Competition has caused pricing wars between some gas stations. Many consumers shop several stores in an effort to buy the same product at the lowest price. Comparative shopping and pricing to get the best buy has become very popular because of the high cost of living.

Buying on sale means that there has been a reduction in price to sell more. Sale prices are reduced either at a fraction off the original price (⅓ off) or a percentage off (cut 10 percent).

Unit pricing of food can help consumers determine the best product value. Large economy sizes are not always the best bargain. Unit pricing tells the consumer how much one unit of the product costs. A box containing 1.81 kg (4 lb) of soap powder may cost $4.69. The unit price would be listed as $2.59 per kg. ($1.17 per lb).

Working in Manufacturing

The jobs are similar in most manufacturing plants, however, the working conditions can be different. Steel workers have a very hot and dangerous working environment. Automobile assemblers and other assemblers are under constant pressure. They must do the same job repeatedly, which can become very monotonous. Electronic workers are considered to have some of the best working conditions in industry. This is due in part to the product and requirement for clean, well-lighted work areas.

Working conditions are affected by safety laws and organized labor unions. Labor unions stress job satisfaction and the employee's well-being to the employer. Employees and their families can receive medical, dental, and optical care as a union

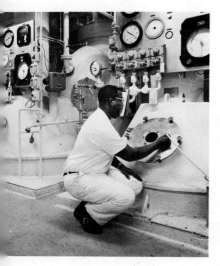
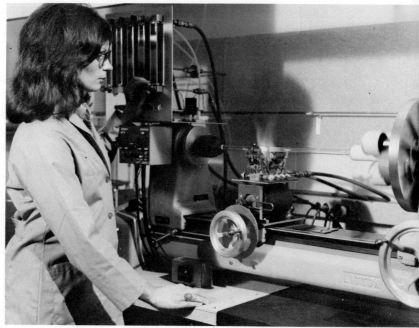

left An employee of a pharmaceutical company. *right* A ceramics engineer making a glass fiber.

benefit. Management, unlike labor, does not have a union to represent their needs. A committee of management employees usually expresses its needs to the company's owners.

Safety laws require some employees to wear safety equipment such as glasses, safety shoes, and safety hats. Safety laws also require employers to provide safe, well-lighted and ventilated working areas. Even with strict safety laws and enforcement, accidents still occur. Employees have accidents every day because of carelessness.

Education and Training

Changes being made in manufacturing are creating increased need for highly skilled workers. As workers are displaced by automation, they need to be retrained for the newer jobs. The workers driving fork lifts or stocking warehouse shelves may be displaced by automated materials handling systems. A person welding auto body panels or assembling televisions may be displaced by robotics systems. These persons need to be retrained to operate the new equipment. If not, they will have to take lower paying jobs or be out of work.

These more technical jobs also require a higher level of education and training. Training beyond high school is often required.

People working in management usually have to have a college degree.

The level of education for each of these job classifications increases from semiskilled to management. High school graduates often fall in the unskilled worker area. Vocational high schools are providing skill-development training. This training could result in a skilled worker. Semi-skilled workers are ones that have been trained to do one job. The training received may be the only formal training the company will provide.

Skilled jobs often require training after high school such as technical school, community college, or apprentice programs. These training programs could give the graduate a technical certificate. These studies include only courses in the field of study. Courses in math and English are not included. Two years of study at a community college, with courses in math and English lead to an associates degree.

Professional jobs require a college degree. In areas of science, engineering, and business, a master's or doctorate degree may be required. A college program of studies would include courses in English, math, sciences, and social sciences. These courses increase in difficulty each year. Although 70 percent of the jobs in manufacturing do not require a college degree, the college graduate usually has a higher income.

A new employee receives on-the-job training.

Methods of Training

Employees are trained sometimes by the company's personnel department. Semi-skilled employees are trained to fill a position such as a machine operator or stock handler. Training may involve classroom work and practical on-the-job training. Orientation of the company's total operation may be a part of the training. Safety training is very important for the employees.

Apprentice training programs begin employees at a higher starting wage. The apprentice must attend adult classes two evenings a week for information about the worker's trade. After three or four years of training and an increase in wages, the apprentice will be tested on his or her knowledge of the trade and, if successful, will be issued a license. The apprentice will then be called a journeyman in such occupations as electricity, plumbing, or printing. Training programs other than on-the-job are contracted by private schools to train employees.

Job Entry

To get a job with a manufacturing company the following steps are common:

1. Submit a letter of application or fill out an application.
2. Submit a résumé of vital information about your background experiences.
3. letters of reference.
4. Interview with personnel manager.
5. Follow-up.

Applications for employment are usually simple to complete, but it is important that all questions be answered correctly. A letter of application is usually mailed to the personnel department to be filed for a possible position. The letter is a way of inquiring about a possible position or opening.

A résumé of past experiences includes childhood, high school, and higher education. Hobbies, organizations, and awards are included to get a picture of the total person.

References from former employers are important. A good reference informs the company that you have a good attendance and safety record. A company will place a lot of trust in a reliable reference.

Most jobs require some type of interview, either formal or informal. Job interviews are conducted between the job seeker, interviewer, and the personnel manager. The interview is a fifteen to thirty minute (or longer) question-and-answer period. The impression left after the interview could be the deciding factor in getting the job.

A job seeker should follow-up after all the paper work and the interview are completed, to make sure the company has every item needed to consider you for the job. Then call back to see if the position has been filled. An interested person will follow-up on a job. A company wants interested employees.

Advancement
Promotions or raises in wage or salary are called *advancements*. Large companies have a well-defined policy about advancements. Under this system employees will know that they are considered for a raise. Every employee should be reviewed and evaluated each year for possible advancement. There are two ways to improve one's position—either a new position higher than the previous one or a raise in pay for the same job. Normally an employee will advance in salary each year because of additional experience gained on the job.

Occupational Opportunities
Background on the job may qualify employees for a better job. This is occupational opportunity. The chance for a promotion because an employee has had several good experiences can make him or her a very valuable employee. The employee who can gain the right experience will have unlimited opportunities.

You—The Consumer of Manufactured Goods

Consumer Agencies
Consumers are the people who purchase the goods and products of industry. We are all consumers of one product or another. The products that are placed on the market should benefit the consumer. However, this is not always true. Some products have been manufactured, marketed, and later found to be harmful. Many toys are considered unsafe because of sharp edges, harmful paint, or loose knobs babies may swallow. To solve some of the problems a public campaign was launched. This led to the forming of the Consumer Affairs Agency in the late 1960s under President Kennedy.

Some companies test products for minimum standards. One such organization is the Underwriters Laboratory. The tests are usually conducted on electrical and construction materials for flame resistance. Next time you make an electrical purchase, look for the (UL) seal of approval. This will assure you the product has met minimum standards. To protect against electrical shock and

To ensure that all the parts of a new camera are in working order, tests are conducted.

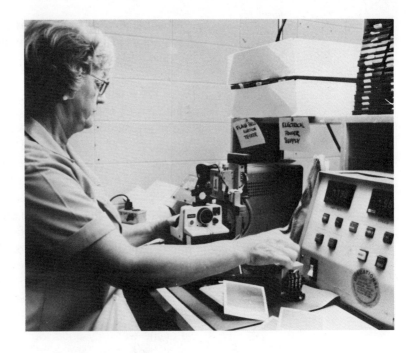

fire hazards, televisions, radios, stereos should carry the (UL) seal.

The consumer agency is working in government to protect the consumer from possible danger from products. The most popular consumer advocate is Ralph Nader. Nader has worked to bring to light problems in automobile safety. As a result, Congress required the auto makers to equip all cars after 1970 with safety seat belts. Later seat harnesses were required. Today, automobiles have warning systems when front seat belts are not fastened. Air bags may be required in autos in the future to prevent injury.

Government Agencies

The Federal Food and Drug Administration is a government agency that controls all new food and medical products to be introduced to the consumer market. A new medicine or food must meet the FDA's standards before it can be released to the consumer market. Some products are passed and later found to be harmful to the health. A commercial sweetner called cyclamate once used to sweeten drinks and foods was banned from the market because it was found to cause cancer in rats. A red food dye used to color cherries and other foods has also been banned for the same reason.

The banning of these items helps assure the consumer that constant studies are being made in an effort to keep harmful products off the market.

Be a Wise Consumer

Being a wise consumer is not easy in today's market place. There are many brands of the same product on the market. How does one decide which ten-speed bicycle to buy? What does one consider in purchasing a car, stereo, radio, or camera? The answers to these questions could help any consumer make a better purchase. A consumer's guide is available to help make the decision on which television or stereo is rated best in performance tests. These tests are conducted on each product to see how well each compares to the others. The results are printed in the *Consumer Report.*

Further information about any product can be obtained from the retailer who sells and services the product. The term *service a product* includes adjustments under warranty and repair service after the warranty ends. Some retailers have created advertisements based on service as a reason to buy from their company.

Service of a product is an important consideration when buying products that might need service regularly. Televisions, automobiles, stereos, and appliances normally need some periodic service. Refrigerators, sinks, stoves, normally need service less often. Some items—wheelbarrows or shovels—do not usually require servicing.

Suggestions for Further Study

Observations

We are surrounded with examples of manufacturing. Try to discover what manufacturing plants are located in your community. What products do they make? How large is the company? How many different products do they manufacture?

Look around you. Can you see examples of the steps in manufacturing? Marketing surveys? Product development? Product testing?

How are products marketed? What kinds of advertising can you see? How are the products distributed?

How can you become a better consumer? Do you check package labels? Do you know what you buy? Are you a smart shopper?

Questions

1. What idea of making parts formed the basis of modern manufacturing?

2. What resources are necessary for producing a product?
3. How does a new idea for a product become a mass production?
4. How can 55 automobiles be made in one hour?
5. Name some methods of getting a job in industry?
6. What groups or agencies protect the consumer from harmful products?

Personal Involvement

You can get involved in manufacturing. One way is to study one manufacturer in depth as a class or individual project:

1. Select an industry interesting to you and your class such as auto, steel, ship, or plastic.
2. Organize the class into a manufacturing corporation with positions such as president, plant manager, and safety engineer.
3. Research job descriptions for each position and have a student play that role in the company project.
4. Write letters. Have each member write a letter requesting information about the industry's operations.
5. Design a product the class will build. An example is a display with each section or process of the production represented. Each section would be constructed to scale and have some lights or explanation of the processes.
6. Divide the display into work units so each student will build a section.
7. Research each process and write letters for more specific information, if needed.
8. Build a display of industry while conducting weekly progress reports from each section of the class project.

Organize the class into labor and management to mass-produce a product. Form a corporation, sell stock, pay wages, sell the product, and divide the profits. Other activities are described in the ACT manual.

Key Words

manufacturing	production
Industrial Revolution	plant layout
interchangeable parts	assembly
assembly line	flow chart
specialization of labor	quality control
finances	testing
human resources	marketing
natural resources	distribution
stockholders	advertising
management	wholesale
design	retail
research and development	consumer
feasibility study	just-in-time manufacturing

Pen Assembly Line

Concept: *Line Assembly*

The use of an assembly line to make products was an innovation of American Industry. Henry Ford used the assembly line to produce automobiles. He reduced the time required to assemble an auto. This reduced labor and made it possible to reduce the selling price of the auto.

Objective

In this activity you will study efficiency in assembling a product. You will compare the time needed to assemble products one-at-a-time with the time used in assembly line organization.

Procedure

Ask your teacher to get fifty inexpensive ball point pens. These are available from businesses in the community who give them away for advertising. The pens should be made of several parts that can be easily disassembled.

First Test Identify five students to conduct the test. If possible use both females and males as workers. This will give you a good cross section of your class.

Disassemble the pens, placing the parts from ten pens in each of five trays. No parts should be assembled.

Set up the test room with a chair and table for each participant. Place the participants around the room so they cannot see each other. A member of the class can act as judge for each of the participants. The rest of the class will be observers. The observers will record the procedure each participant uses to assemble the pens.

Select a timekeeper for each of the workers. This person will record the time required to assemble the ten pens.

Place the trays of pen parts on each worker's table. The workers are to assemble the ten pens from start to finish. Time each worker to determine how long it takes to do the work. Record the following information:
- name of worker
- time required to assemble all ten pens
- order of assembly (first part assembled to last part assembled)
- how parts were organized before test began
- do all pens work? (add 30 seconds to time for each pen that does not work)

Evaluation

At the completion of the test, the class should analyze the procedure and results. Could the workers have done the work more efficiently if they had practiced their procedure? Who assembled the pens in the shortest time? What procedures did this person use? Could other workers improve their time using this person's procedures? Brainstorm new procedures for assembling the pens in the most efficient way.

Second Test Disassemble all the pens. This time place all like parts in trays. You should have a tray for each part: spring, barrel, cap, etc. Place the trays in a line on a long table. This table will be your assembly line. Place the trays in order of assembly, from left to right.

Select a new group of workers to assemble the pens on the assembly line. There should be one worker less than the number of parts in the pen i.e., a pen with eight parts would require seven workers). The first person puts the first two parts together. The subassembly is passed to the right to the second worker who adds the third part. The next worker adds the fourth part, and so on until the pen is complete.

This assembly of one pen can be used as training for the assembly line workers. Disassemble the pen and put the parts in the proper bins.

Select a person to be timekeeper. This person will record the time required to assemble all fifty pens.

When everyone is ready, begin assembling the pens along the assembly line. Continue until all fifty pens are complete.

Evaluation

How long was needed to assemble the pens? Calculate the average time per worker. How did this compare with the time per worker when the pens were assembled one at a time? Could you have improved the efficiency of the assembly line?

When you set up an assembly line in your study of manufacturing, this experience should help you do it more efficiently.

In modular construction, large sections of a building are prefabricated and then assembled on the site.

Chapter 5   Construction

Introduction

People have been building shelters to protect themselves from the environment for 8,000 years. The shelters or buildings they built can give us a key to understanding the technology of construction. Homes were made of materials that were available to the builders. The Egyptians gathered mud and straw to form brick. Two ton limestone blocks were also used by Egyptians to build the great pyramids. Just how these pyramids were built is still unknown.

The base of the great pyramids covers almost five hectares (50,000 square meters) standing 147 meters (485 feet) high. Around 2500 B.C., slaves were used to move the limestones to the site, which was kilometers away. One Egyptologist has estimated that it took twenty years and 400,000 slaves to build this pyramid. Pyramids have been visited by the leaders of many nations who marvel at the sight. History marks the pyramids as one of the Seven Wonders of the World.

Greeks built temples to worship their gods. One of the most recognized temples of this period is the Parthenon in Athens, Greece. This great building was built around 400 B.C. The earliest Greek theaters were built around 450 B.C.

Roman architecture was copied from the Greeks. Rome's Triumphal Arch is a landmark to the buildings of that time. Romans were the first to use concrete in buildings. From the arch a dome was developed by the Roman architects. The dome formed the basic structure for famous temples and later for churches.

top left Arch of Constantine, Rome. *top right* The Pantheon, Rome. *bottom left* Eiffel Tower, Paris. *bottom right* Modern steel frame building under construction.

After the decline of Rome, changes in architecture occurred mostly in northeastern Europe. Great wars influenced the architecture of the period: castles and fortresses were the main buildings of the time.

Around the fourteenth century French artists rediscovered the Roman arch and the Greek design. Paris, France, has several famous buildings that reflect the Roman-Greek Arch design. The rulers of France developed these buildings as a tribute to the greatness of their country.

Many new building materials were developed over the next four centuries, changing construction methods for all time. Iron, coal, and steel made it possible to erect a skeleton structure. This made buildings much lighter and stronger. In 1890, a steel structure was built in Paris that was a marvel to the world. The Eiffel Tower was built to reach over 300 meters (990 feet) in the air. Iron and steel manufacturing in Europe made this possible.

Today, skyscrapers like the Empire State Building in New York and the Sears Tower in Chicago are made possible through the use of steel. Thirty-, forty-, and fifty-floor buildings are made of steel skeletons and covered with walls of glass. This adds beauty to the skyline. These buildings will remain for generations to come. The period in which we live will be viewed by future generations based largely upon these buildings.

Dwellings

When early people lived in caves, there were no houses because there were no tools to cut and shape materials. The only shelter for these people were caves and naturally formed dwellings.

Cave dwellers made tools of stone and sticks. Later people learned to break the soil for planting with a plow. The plow may have given people ideas for other tools that could be made for cutting trees. Trees and branches were used to make one-room huts. Clay was also a popular building material. Bricks made of clay were used to build shelters.

When crops were good, extra food was traded for family needs. Soon people began working to meet those needs. Various trades developed like that of a baker, cobbler, or carpenter. These people settled wherever there was a need for their service. Villages were formed where fertile land and water could support their needs. Animals were trained and domesticated to work the land and provide food.

Villages developed into cities. The ancient city of Priene, Greece, was one of the first. This city had a stone wall completely surrounding it for protection from enemies. Houses were small with one room. This one room often housed an entire family.

In Athens, an ancient Greek city, stone was the principal building material, and stone cutting was an art form. This artistry can be seen in the statues of the Greek gods.

Tools made of iron made possible the use of new construction materials. These tools also made the cutting and shaping of lumber easier. In areas where forests were available, wood became an important building material. Lumber was used in Europe and America to build wood frame houses. A frontier house of the 1600s was built of fallen trees notched on the ends. The early family lived in this log cabin. Today the lumber industry supplies the material needed to meet the ever increasing need for housing. Wood is a lightweight and strong material. It is also less expensive than some materials, such as stone.

Brick buildings were as popular in colonial Williamsburg, Virginia, as they are today. Brick construction is more expensive than wood because of the high cost of labor to lay the brick. A tall brick building would cost a great deal for this reason.

Building materials have changed from sod to tree limbs, to stone, to lumber, to iron, and many other modern materials. These changes have come about as a result of improved technology. Changes in construction are easily observed in the building materials of today. The changes have also affected building methods. Plywood, a modern building product, is an example. Its uses are unlimited in building. Forms of plywood are found in flooring, exterior and interior walls, roofing, cabinets, for example.

Transportation

Transportation of goods and people are provided through construction. Roads were constructed both for personal travel and for vehicles that haul food and a variety of other goods and products. Airports were built for the landing and taking off of airplanes. Railways and terminals were constructed for mass transportation. Canals were built for ships to use. These examples show how transportation depends upon construction. As you will see, transportation has made progress possible through advances in construction.

Land—Roads

The need for roads came soon after the discovery of the wheel. Roads were first built by Egyptians and Romans. The Romans are known for their road-building technology. The early Romans

HISTORICAL DEVELOPMENT OF CONSTRUCTION

Time	Development
2500 B.C.	Construction of the Pyramids, Egypt
123 A.D.	Pantheon built in Rome
1643	Taj Mahal completed in India
1851	Crystal Palace completed in London
1883	Brooklyn Bridge opened—5,989 feet long
1885	First structural steel beams rolled
1889	Rand McNally Building constructed using steel columns throughout—Chicago
1889	Eiffel Tower completed in Paris, France
1903	First skyscraper (16 stories high) to have a reinforced concrete framework, Cincinatti, Ohio
1908	Singer Building completed, New York City
1912	Long Key viaduct opened—linked Key West, Florida with the main land
1924	First concrete shell roof constructed—used as a roof for a planetarium—Jena, Germany
1926	Transit mixer—concrete mixer on a truck—developed. Becomes indispensable piece of construction equipment
1928	Rove Tunnel completed, France. Canal tunnel connecting Marseilles to the River Rhone
1931	George Washington Bridge opens connecting New York City and New Jersey
1936	San Francisco Bay Bridge opens

continued

needed roads to move their armies and equipment. What the Romans knew about road building was surprisingly advanced.

The Roman road was constructed so rain water would drain from it. A ditch was dug alongside the road to carry the water away. The base of the road was important and built in layers. The first layer was composed of material from stone quarries and rubbish from the cities. Flat stone set in mortar was next, followed by concrete and crushed stone. The top layer was flat paving stone. A curb stone was used to keep the edge from wearing away. The thickness of such a road varied. The Romans built 80,000 kilometers of roads. Many of these roads still are in use.

The Roman army built temporary bridges out of many boats. These bridges, called pontoon bridges, moved armies across rivers. This idea is used by modern armies. Roman cities required large bridges made of stone. Some are 52 meters (171 feet) high and 2,000 years old.

The Romans had many slaves to do the work. The goal was to build a road for quick transportation of people and equipment. The Roman road may be compared to the roads of today. As in Roman times, modern roads are built in a straight line. Concrete and crushed stone are used for the roadways. Today, however, steel is used to make the concrete stronger, and large earth-moving equipment has replaced the slaves. Modern highways are constructed using the same layer method. The modern interstate highways are the result of this progress.

Railroads have improved with advances in construction techniques. The development of the railroad began in England's coal mines. A muddy and rough mine road made it difficult to remove the coal. Wooden rails were set over the roadways. Wagons with wooden wheels were pulled over the rails. Horses and men could easily pull wagons loaded with coal. Around 1750 the fragile wooden rails were replaced with iron ones. This gave increased strength, and the rails did not wear out as fast. The wagon wheels were designed with a flange to keep the wheel on the rail.

Railroad Construction Method

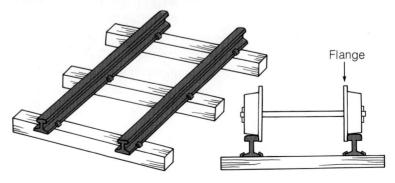

HISTORICAL DEVELOPMENT OF CONSTRUCTION

Time	Development
1940	Tacoma Narrows suspension bridge over Puget Sound opens and 4 months later collapses, caused by wind-induced vibration
1941	Grand Coulee Dam opens — hydroelectric
1944	Delaware Aqueduct completed — 105 miles long — supplies water to New York City
1967	Largest hydroelectric power station in world constructed in Siberia
1974	Sears Tower constructed — 110 stories high, structural steel framing

The railroad developed as a result of a number of technical advances. Some of these were advances in construction:
1. Roadbeds constructed of stone to support heavy loads
2. Iron rails manufactured to standard size
3. Development of an energy source
4. Invention of the airbrake to slow and stop the train
5. Quick coupling and uncoupling of train cars
6. Construction of train bridges across rivers and valleys
7. Tunnel construction through mountains
8. Switching circuits to route trains
9. Signals to operate more than one train per track.

Rail transporation today is economical for hauling coal, ore, new automobiles, and other cargo. The construction industry played an important part in the development of railroads.

Waterways

The need for transporting goods increased with factory production. Long canals were dug in the early 1800s and became major trade routes. Early canals were dug as a part of a major river with some running great distances. Horses and mules were used to pull the flat-bottom boats through the canals. This method of travel was expensive to construct, and traveling time was slow. Expanding the canal system was difficult and was stopped after the railroads began in the 1850s.

However, in America the canal system did expand in New York and around the Great Lakes in the early 1900s. Canals were combined to form major shipping routes. The Suez and Panama Canals are two famous canals of high value. The Panama Canal, 81 kilometers long, links two oceans, the Atlantic and Pacific. The Suez Canal links the Mediterranean Sea with the Red Sea. It is about 160 kilometers (96 miles) long and shortens the route between England and India by 9,600 kilometers (5,760 miles).

Construction of the Panama Canal first began in 1882. Several countries tried to dig the canal before it was finally completed in 1914 by the United States at a cost of $380 million. In 1913 there were more than 43,000 workers working on the canal. An estimated 160 million cubic meters (208 million cubic yards) of land was moved. This was done by steam shovels and dredging operations. Cuts were made through hills and jungles. Twenty-three million cubic yards of earth were used to dam a river in the area. Three sets of locks were constructed to raise the water level in order to permit large ships to pass through the canal.

A modern airport is a major construction project.

Air Travel

Two bicycle mechanics began air travel. The Wright brothers began building models in 1899. Having succeeded in unpowered flights, the brothers wanted to build a powered glider. They designed and built a 12-horse power (9 kilowatt) engine. A propeller also was designed and tested. On the morning of December 17, 1903 the first power-controlled flight was made by Orville Wright. The flight lasted only 12 seconds covering a distance of 37 meters (122 feet). The plane was launched from a single-rail catapult.

Modern airplanes require a runway for take-off and landing. The construction of early runways was simple. Any long stretch of land would work. At the end of World War II, mass travel on airplanes began, and many of the military transport planes were used to start airlines.

Today airports are major construction projects. Some old airports are rebuilt to meet increased traffic. Airport locations and operations are controlled by the Federal Aviation Administration (FAA). The FAA sets standards for building airports. The size of an airport depends on the service to be offered. A large airport handling jet passenger planes needs 250 to 600 hectares (625 to 1,500 acres) of land. This is much larger than a train or bus station.

Construction

The location of the airport is very important and sometimes poses problems. Where could a city find land enough to build an airport? Farm land or land near the ocean is usually first choice. The noise of airplanes flying over homes in populated areas causes complaints. But an airport should not be too far from a city. Airplane passengers want to be close to the downtown area.

An airport consists of such things as a terminal building, control tower, runways, and parking terminal. The terminal building contains ticket counters; baggage checks; waiting, arriving, and departing areas; restaurants; shops; and offices. This building is usually very large and modern. It is designed to enable planes to load and unload passengers at the terminal. The terminal must also provide public transportation such as buses, subways, taxies, air taxies, and elevators.

Only a large construction company can build an airport. The company must have a full staff of employees and equipment. Experienced airport builders usually go from one airport job to the next. The total construction time of an airport can be five to ten years.

A control tower is a relatively simple building. The location is important, and the equipment is even more important. Air-traffic controllers must be able to see all planes on runways and taxiways.

Runway construction is similar to road construction. A firm base of sand, gravel, and soil, and a layer of reinforced concrete serves as the runway. Concrete is the best material for the job. Large, heavy planes must take off and land on this runway.

Runways are very long and wide. A jet airliner must have a runway 4,000 meters long (13,200 feet). The ends of the runway must be free of trees and buildings. This area is called the *clear zone*. This zone is for planes to climb and descend. A distance of 800 meters (26,400 feet) of clear zone is required for large planes.

Public transportation to and from the airport is a major concern. Passengers arriving at and departing from the airport must have transportation services available. Most airports have two separate areas: one for arriving passengers and one for those departing. This arrangement reduces traffic problems. Construction of mass-transit services is a part of large airports. A mass-transit company might construct a rail line to the airport. Passengers are directed to the system by signs. More than 1,000 planes may land and 50,000 passengers may arrive and depart in one day.

Impact of Construction

If you stop to look around, you can see the effects of construction. The homes in which we live are strong and well built. The schools we attend are designed for learning. The roads on which we ride are smooth, straight, and safe. Shopping centers contain many shops, all under one roof. Civic centers provide a place for entertainment such as athletic events, music shows, and even circuses. All of this and more is made possible by construction techniques. These buildings and roads are built using the latest construction methods.

What effect does modern construction have upon our lives? Do we live better as a result of a super highway? How do we value the shopping center around the corner? The answer to these questions is different for each individual. Some people will move away from an area that has become highly developed. Others would move into the area because of the convenience and services. Construction projects are usually built to satisfy a need. The needs of a community are being met by companies that build houses, apartments, shopping centers, and roads. Where there is a need strong enough, construction personnel work to meet that need.

These photos show the same area in Hartford, Connecticut. *left* Before the buildings were torn down. *right* Buildings constructed on the site.

Many new homes are being constructed with solar panels for collecting the energy of the sun.

Energy shortages have caused changes in construction. Homes and buildings must be better insulated. Some homes now are designed to be heated by solar heat collectors. Old homes may be converted to solar heat. How will the construction industry meet these needs? The gap between home construction of yesterday and the solar-heated house of tomorrow must be bridged. To do this, new technology must be developed. A new type of engineer, the solar engineer, is being trained. Construction companies will employ these solar engineers to design solar homes and buildings. New materials will be required and workers will be trained to use them.

The use of solar energy is creating a new era in construction. The changes in construction will begin slowly, then increase rapidly. Every effort to meet this need will be made by the construction industry.

Technical Advances in Construction

Tools and Machines of Construction

Many tools were developed by the Stone Age people around 8000 B.C. Basic tools made of stone were in use around 4000 B.C. When metal was discovered around 3500 B.C., tools were shaped from metal. Egyptians and Romans were the first people to have tools made of iron.

Iron was in limited use, so many people still farmed with wooden plows. Tools and weapons made of iron became more common around 1400 A.D. When the Spanish and English colonized America, iron saws were in use. A saw mill was built in America about 1615.

Before the development of saw mills, lumber was cut by people using long saws. Two men would pull the saw back and forth cutting the board from the log. More boards were cut, one at a time.

In early saw mills similar saw blades were used. They were powered by the large water wheel outside the mill. The circular motion of the water wheel was changed to an up-and-down motion. This changing the direction of motion in machines was a great technological advancement. By using several saw blades fastened together, a log could be cut into several boards at one time. A larger number of boards could be cut than could be cut by hand.

An increase in the amount of lumber made more construction possible. Houses, stores, churches, and other buildings were built. The people who built the homes were called *carpenters*. Communities depended upon them for their ability to work with tools and wood. A carpenter was a very important person to the community. Without a carpenter, the community could not grow.

Later, circular (round) saw blades were developed. The circular saw blade appeared in America in the early 1800s. The circular saw blade made continuous cutting possible. Instead of moving up and down where the blade cut only on the down stroke, the circular blade was always cutting. Lumber could be cut much faster.

Table Saw

The circular saw used in the saw mills was made smaller. The smaller blade was mounted under a table. Thus, the table saw was developed. The early table saws used in mills were powered by the large mill wheel, usually through a system of belts and pulleys. Electricity and electric motors converted the water-driven saw into a modern tool of construction.

Today electric motors provide power to the tools of construction. The most commonly used saw in construction is the portable saw. This saw is used by carpenters instead of the hand saw. The builder's saw is a more accurate electric saw that makes easier difficult cuts on large lumber. Other hand electric tools improve construction work.

These huge saw blades are operated by means of a computer.

A radial arm saw is another standard construction tool. It was developed by Raymond Dewalt in 1922. The saw blade is mounted over the table on an arm. Many sawing operations can be performed with this saw. Sawing lumber to the right length is an easy operation with this saw.

Modern electric-powered tools have improved the construction job. Cordless power tools can also be used on any construction site. More accurate work and the use of new materials are results of improvements. Labor saving is another advantage. The hard labor of digging a drainage ditch has been eased today by a backhoe that does the work of five laborers.

Materials of Construction

The materials used in construction have changed little over the years. Wood, brick, and stone are still the most common. However, tool technology has changed the way carpenters and masons work with these materials. Manufacturing technology has improved and increased the number of wood and masonry products. However, the raw materials of wood and masonry are the same today as they were thousands of years ago.

Wood. Efforts to use all parts of a natural resource such as the tree to make products has been a benefit to every consumer. As wood prices increase, the need for cheaper building materials

Logs in a lumber mill are cut into rough boards.

also has increased. As a result, many wood products are made from wood chips and glue to reduce cost. Particle board, vinyl-coated paneling, and construction lumber are examples of these products. Use of the total tree is also a way of conserving wood. Through technology we are learning how to get the most from a tree.

Bricks. Bricks are made from natural clay deposits in the earth. The Egyptians made brick of clay and straw. Baking the brick causes the clay to bond, holding the shape. Modern bricks are quite different from the early sun-baked bricks. Bricks now are uniform in size and strength. The outside edges can be coated with a glaze to make them any color. Brick has been used throughout our history. The shape and material has not changed a great deal. The brickmason's trade is an equally old and honored profession.

    New building materials have changed the tools and the technology of building. Have you ever wondered why a building over forty stories high does not fall over when the wind blows? We have all tried stacking wooden blocks on top of each other to build the tallest tower. The desire to build the tallest building is found also in people who design buildings—architects. The development of new materials and techniques have helped architects reach their goals.

Concrete and steel form the base of a new building.

Concrete. Concrete, steel, and glass are modern construction materials. These materials have made it possible to construct tall buildings, long bridges, and durable roads. Concrete is used to build foundations for homes, office buildings, and bridges. The foundations of buildings and bridges anchor them in place. *Concrete* is perhaps the most widely used construction material today. Concrete begins with a compound made by burning a mixture of limestone and clay. The burned mixture is then ground to a very fine powder called *cement.* A chemical reaction causes cement to harden when water is added. When the cement is mixed with sand or gravel and water added, concrete is made. The reason concrete is strong is that simple cement surrounds the material and bonds it together on all edges.

Aluminum. Aluminum is a very popular material made from bauxite ore. Aluminum siding on houses is less expensive than brick. It has a baked-on color and does not rust. These qualities make aluminum siding a good construction material.

Aluminum is also used in office buildings because of its light weight and bright finish. The mobile home industry is a big user of aluminum. Mobile homes are covered with aluminum siding to reduce weight. Aluminum has many possible uses in the construction industry. Wherever aluminum is used, the homeowner benefits. The construction worker also benefits because of its ease in handling and forming. The many uses of aluminum around the house and school might surprise you. Here are a few uses of aluminum you can look for:

Storm doors and windows	Window frames
Siding	Screens
Sliding door frames	Ventilators
Gutters and downspouts	Boxes
Electrical wire	Furniture
Electrical connections	Nails
Cookware	

Steel. Steel is used in many large buildings. It forms the "skeleton" around which the rest of the building is built. A steel framework is erected on a concrete foundation. This supports the floors, walls, ceilings and other parts of the building.

Glass. Glass has been used for many years in windows. Today it is also used for walls. The outsides of some buildings are nearly all glass. The large pieces of glass can be made very strong. Tinted and coated glass can be used to control the radiant solar energy.

Modular construction can speed the building process. Homes can be ordered from a catalog. They are manufactured in two parts for a one-story home or four parts for a two-story home, and then trucked to the site. Erection on the site takes a short time. This module idea also is being used in apartment buildings.

Technology is helping the construction industry grow. New materials means new ideas for their use. The changing trends in construction do force the construction companies to keep up to date. This updating can be expensive. New materials often require new machines for cutting and installing.

A roof is lowered into position on this modular home.

Construction to Meet Human Needs

Land Development

When someone owns land it is often referred to as his or her *property*. Land is bought and sold through a legal action. Counties and cities keep an accurate record of who owns land within their boundaries. This is for tax purposes. A *title* is issued to the landowner by the county or city land record office. The title is legal proof of ownership. The records of who has owned the land are also kept on file for hundreds of years. This information is studied by a lawyer. A lawyer searches the record book to make sure all sales of this land for the past one hundred years are correct and complete. This process protects the land owner from any previous owner's claim to the land.

To identify the land, a legal description is made. This description is written in legal terms and is called a *deed*.

> All of lot 32 Section 9-B Dale City as the same is duly dedicated, platted and recorded of said County in Deed Book 669, at page 259.
> S 50 46' 00" E, 82.87'
> N 39 14' 00" E, 30.33'
> S 35 50' 18" W, 30.98'
> N 50 46' 00" W, 85.00'

On the deed distances are described in feet or meters and direction in degrees. This is necessary to describe accurately the piece of land. Suppose you wanted to build a fence around your land. Without an accurate description, the fence might be built on your neighbor's land. It could cost the landowner a great deal of money to move the fence. If a building were built with a corner on someone else's land, it would be called *encroachment*. Land is surveyed to prevent encroachment. *Surveying*, determining where the boundary lines are, is done by a registered land surveyor.

Survey parties require a team of workers and equipment. The instrument used to measure angles is called a *transit*. The operator is responsible for the reading and the instrument. Other members of the team measure the distances, post markers, and take notes.

Land is also described by its ups and downs. This is called *elevation*. The term "sea level" is used to describe land elevation in relation to the level of the sea. Any increase above sea level is measured in meters. Elevation is a useful measurement. Land description by elevation will identify valleys, mountains, and flat land. A person who plans to buy land can study this description.

Members of the surveying team sight the next point.

An elevation survey is also called a *topographical* survey. A topographic map is a general map of large or medium scale, showing all important features, including relief. To complete a survey the land must be measured into squares. The size of the square determines the detail of survey. For example, a contour map of land divided into three meter (10 foot) squares is more detailed than a map divided into sixteen or thirty (50 or 100 feet) meter squares. One way of surveying contours is to take an elevation reading at the corners of each square. These figures are recorded on the drawing of the land. This drawing is usually a plot map. After all corners are measured, the range of elevation is studied. The study will determine the elevation lines. The elevation lines are called contour lines. The more contour lines drawn, the more detailed the map. A map with contour lines every three to five meters (10 to 15 feet) would be very detailed.

Each contour line is an elevation. This would be the distance above sea level. From the lines one can tell the following:

Lines	Description
Close together	Steep hill or mountain
Far Apart	Flat land
Circle	Top of hill or mountain or body of water
Crossing	Overhanging cliff
Evenly spaced	Gradual incline/decline

Contour maps are difficult to understand at first. But, after studying the lines, one can get a mental picture of the land's surface.

right Contour map. *left* A relief drawing of the same area shown on the contour map.

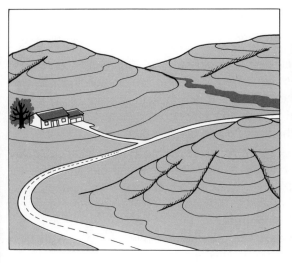
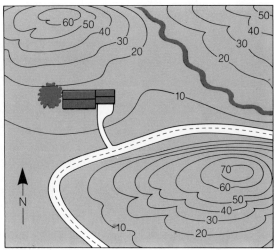

Land Testing

Land or soil must be tested before it is used for a building site. One test is for its ability to soak up water. The test is called a *percolation* test. Land must pass this test before a permit to build is issued by the city or county. Soil with too much water content would not be stable enough to support a building.

Soil also is tested for stability, compaction, and location of rock. Which of these tests are made depends on the type of building or structure planned. A bridge site would need all the above tests before construction. A house site would need only a percolation test. A multistory building site would receive all the tests.

Selecting Land

The location of the land is very important. This is true for any project—house, business buildings or civic building. Several factors affect the purchase of land:

Factors	*Example*
Utilities	Gas, electricity, and water availability
Price	Value of land
Location	Nearness to schools, shopping or recreation
Soil characteristics	Will or will not support building
Labor supply	Availability of workers
Weather	Climate and temperature
Transportation	Availability of buses, roads, trains, or airports

One should consider these factors and then decide on the best site for the price. This is what land selection means.

Clearing Land

Land must be cleared to begin building. This can be done with several pieces of equipment. Trees, rocks, old buildings, or anything else on the building site are moved. A bulldozer and front loader are used to load and move objects on the site. Once the site is cleared the building can be "laid out."

Planning of the site work can save time. Bad weather can delay actual work. A building site can become one big mud hole for weeks. Proper planning could help prevent this from happening.

Planning and Designing

Planning and designing are the beginning of a construction project. Planning is a process that starts with an idea. To better understand the planning process, let's use an example. The example will be the building of a house for Mr. and Mrs. C. W. Wilson and family. The house is being planned by a professional architect.

Planning Step	Purpose
Study different styles	To select an exterior design
Study different house plans	To select a floor plan
Analyze the plan for traffic flow pattern, living pattern, eating pattern and washing pattern	To determine the Wilson's living style
Select a plan or develop a new one	To fit the exterior to the floor plan
Select a building lot	
Develop the working drawing of the house, which includes site plan, elevation of all sides, floor plan, structural details (roof plan) and schedules of such items as doors and windows	To get list of materials for bidding

Some steps may be repeated. The floor plan may have to be changed. The importance of planning is to study the plans, then make necessary changes. This process is repeated until the plan is correct. Any changes after the house is under construction will be more costly than if they had been made during planning.

An architect usually is employed to help in the planning process. A full-service architectural firm will provide help with or complete the following steps:
1. Supply sample house plans for study.
2. Analyze family needs.
3. Make a complete set of drawings.
4. Recommend a builder or advertise the plans for bid.
5. Work with the builder to make sure all material and specifications in the plan are correct.
6. Aid in selecting a building lot.
7. Other services, according to the size of the firm.

An architect's fee for these services is 6 percent of the cost of the house. For example, if the house costs $50,000, the fee would be 6 percent of $50,000, which would be $3,000. Many large architectural firms will provide these same services. Some custom builders will provide many of the architect's services. One example is having a selection of completed house plans supplied by the custom builder.

Architects and engineers are needed to design buildings and bridges. The same services are required to build any structure. A different method of selecting a builder may be necessary for large buildings. Large buildings require a bidding process. The working drawings and specifications are sent to qualified builders. Each builder has an opportunity to estimate the total cost of the building. This estimate is called the *bid*. Usually the low bidder is awarded the contract to build the structure.

Building is done in several steps:
1. Site work
2. Setting the foundation
3. Framing, roofing, and utilities
4. Exterior and interior finish
5. Landscaping.

Site Work

Let's examine each of the steps and find out more about the process. Land development, as was discussed earlier, includes clearing the site. At this point the land is undeveloped. The site is prepared after a contract is made to build. Site work includes:
1. Surveying
2. Grading building site level
3. Grading entrance to lot
4. Digging a trench for the footing and foundation
5. Digging a ditch for drainage of water.

The architect's drawing of the site is used for each of these.

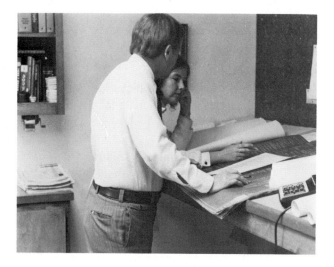

left The architect goes over the floor plans with the customer. *right* A foundation wall on a concrete footing.

Exploring Technology

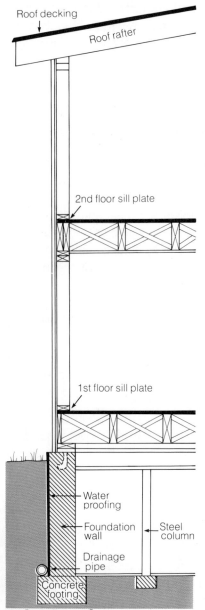

Side View of a Two-Story House

Foundation

When the site has been prepared, the footings are poured. This means that concrete is poured in the trench. The trench is about a half-meter (1½ feet) wide. It goes around the entire building. The concrete is worked into place and is leveled. This will provide a firm level base on which to build.

The concrete footing must harden for several days. The exact time depends on the type of concrete. This process is called *curing*. After curing, the foundation wall is laid directly on the footing.

The foundation wall may be of masonry blocks. The height of the wall will vary. Some houses or buildings will have space under the building. This basement or lower level is partially underground.

In certain areas of the country a basement cannot be built. The land close to the ocean often is too moist below the surface. Houses and buildings in these areas have a crawl space between the ground and floor of the living area or are constructed on a concrete slab.

Foundation walls stop at the floor of the first level. Let's look at a side view of a wall section of a house. Note the parts we have discussed. Other parts will be explained as we finish the building.

Waterproofing the foundation wall is necessary when a basement is included. The foundation wall must be coated with a water proofing material such as a tar product. Drainage pipes and crushed stone are also used at the base of the foundation wall. This carries water away from the house. After waterproofing, the soil around the building can be pushed back. This process is called *backfilling*.

Framing—Utilities

Framing the building can be done several ways. The method described here is wood framing. Some buildings are framed of steel while others are all masonry block. Different areas of the country use different methods. The cost of materials and climate conditions determine the method used. For example, Florida has very few wood frame houses. Multistory buildings require steel framing. The steel framing is necessary to support the floors above. The framing is done by a special crew of carpenters. They work with saws, hammers, framing squares, and so forth.

Anchor bolts are placed into the top row (course) of masonry block. These anchor bolts are used to hold the sill plate to the wall. All framing is fastened to this wooden sill. Next, the headers are attached to the sill. The header holds the floor joists in place. The

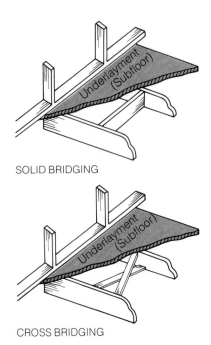

SOLID BRIDGING

CROSS BRIDGING

floor joists rest on the sill plate and are fastened to the header. The other end of the floor joist rests on the girder. The girder is made of either three pieces of wood or a steel *I* beam. This girder supports the middle of the house while the foundation supports the edges. There is only one girder in an average house. The girder is supported by steel posts on a concrete footing.

Floor joists are four to five meters (13 to 16 feet) long and spaced approximately one-half meter (16 inches) apart. Joists are doubled where greater support is needed. Two joists are required to span a width of eight to ten meters (26 to 33 feet). Joists must be bridged together to prevent twisting. Bridging can be solid or crossed.

Subflooring is applied to the joists and headers. Plywood is usually applied first. It may be glued and nailed or just nailed. An insulating material such as tar paper may be used next. The final floor covering will be different in various parts of the house. Ceramic tile in the bathrooms, carpet in bedrooms, and hardwood floors in the family room are some possibilities. However, the final floor covering is not completed at this time. Finishing the floors is one of the last jobs.

Wall construction begins with the outside walls. They are framed with *studs*, which are boards 5 cm x 10 cm (2 x 4's), 2.44

Window and Door Framing

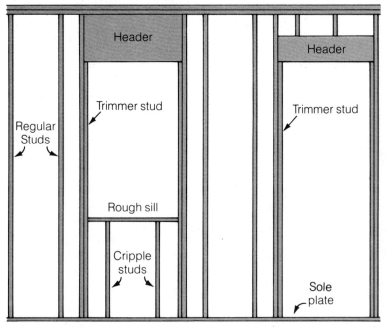

meters (8 feet) long. A wall is measured and the floor is marked. These marks show the carpenter where the doors and windows are to be placed. Areas over windows and doors are strengthened with double studs. These also are called *headers*.

Shortened studs are used under the windows. They are called *crippled studs*.

The interior walls are measured according to the floor plan. The plan will show where each room is located and its size. The walls are nailed together on the floor then put into place.

The roof is built next. This is done to protect the interior from the weather. The design of the roof is very important. In northern areas the roofs are made with a steeper pitch to prevent the build-up of snow. Its shape affects the style of the house.

Roofs also are described by their *pitch*, or slope. A 12/12 roof is very steep, whereas a 3/12 roof has a gradual slope. These two numbers represent the ratio of the height of the roof to the span or length. The first number is the height, the second the span. Each style must be framed differently. The basic process will be described next. The supporting members are called *rafters* and *ceiling joists*. Rafters are measured according to the roof pitch or slope of the house. A useful tool for this is the framing square.

A manufactured roof truss is often used to speed construction. The truss is made of smaller lumber but has more pieces. Its strength is equal to the rafter. Trusses are spaced one-half to three-fifths meter (16 to 24 inches) apart.

Roof Construction

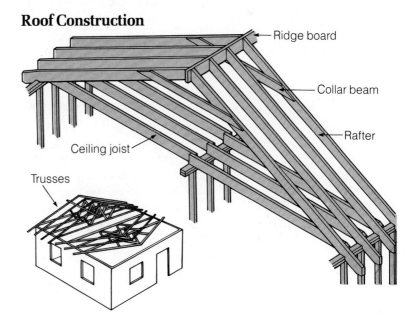

This truss system will supply the structural strength to carry the weight of the roof.

Manufactured trusses have many uses in construction. The most obvious are roof trusses used in place of "stick-built". ceiling joists and rafters. Floor trusses are often used instead of heavy beam joists.

Trusses offer several important advantages over more traditional methods of construction. They can be used over long spans without supporting posts. This allows, for example, for a basement without posts. This is a very important feature for storage buildings. The clear floor allows more efficient use of the space.

Trusses and pre-made wall sections are built in a factory and carried to the building site. This shortens the time needed to build a house. Once the foundation has been laid, construction of the house can be done very quickly. Floor trusses are put in place and sub-flooring put down. Pre-made wall sections are erected. Often these are delivered to the building site and set in place as they are taken from the delivery truck.

Roof trusses are set in place, again, as they are taken from the truck. Roof sheathing is laid next. With the application of roofing materials, the house is under roof, often in a matter of only a few days after the foundation was laid.

Houses built of trusses and pre-made wall sections can be erected in less than one-third of the time needed for conventional "stick" construction. This results in a large savings in worker time and costs. This method of construction, often called panelized construction, is growing in residential and commercial construction.

Once the roof is framed, plywood sheeting is added. The plywood holds the trusses parallel to each other. A weatherproof tar paper is used to cover the sheeting. Over this paper, asphalt shingles are applied.

Shingles come in many colors and materials. They are fastened with nails. Shingles are nailed one row at a time. The first row is double thickness and goes on the lowest point of the roof. Shingles are purchased by the bundle. One bundle of asphalt shingles will cover about 3.6 square meters (39 square feet) of roof.

Shingle Roofing

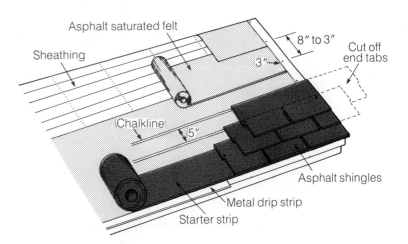

Utilities are "roughed-in." This means wiring is pulled and electrical boxes are installed. This also must be done with the plumbing. All pipes to baths and kitchen are installed. Drainage pipes are also installed. This work is done by electricians and plumbers. Air conditioning and heating workers install pipes inside the interior wall. All of these workers must work to install materials

Electrical and plumbing utilities are roughed in.

before the walls are covered with plaster, gypsum board or wood paneling.

The final work of installing such items as lights, switches, and sinks does not take place until the walls are enclosed. These and other features of the house are inspected by building inspectors. Inspection insures that proper materials are used and installed correctly. This work must be done to meet a local building code. Inspections are required on all construction over a certain value. The value is set by each community.

While the electricians and plumbers are working, the carpenters are hanging windows and doors. The windows and doors are manufactured in a plant. They are purchased in standard sizes.

Once the doors and windows are installed, the exterior finish is applied. The exterior can be finished many different ways. Brick, aluminum, lapsiding, paneling, or a combination of these may be used. Each material requires special workers. Brickmasons lay the bricks to cover the outside walls. Aluminum siding is installed by trained workers. Carpenters finish the trim around the roof.

The area under the edge of the roof is called the *soffit*. The soffit covers the roof trusses and provides ventilation in the attic.

While the outside is being finished, interior work can begin. Insulating a house properly is very important to prevent heat loss. Areas to be insulated are:

Ceiling or Attic	20 centimeters (8 inches) of insulation
Floors	20 centimeters (8 inches) of insulation
Walls	10 to 15 centimeters (4 to 6 inches) of insulation (Depends on wall thickness.)

Wall insulation is installed in the attic.

An insulating material also is used between the exterior siding and the wall. Insulation may be improved by using special windows and doors. Thermal windows have two layers of glass with space between them. Doors are made of metal or wood and filled with insulating materials. Storm windows and doors are also installed to cut down on heat loss. The greatest heat loss can be through ceilings, so the greatest amount of insulation must be placed there.

Finishing

The interior walls now can be enclosed and finished with many different materials. Plaster, gypsum board, and wood paneling are some examples. These materials are fastened to the wall studs with nails or glue. Plaster and gypsum board must be sanded and painted. Ceramic tile can be used in bathrooms to cover walls and floors. Tile is waterproof and easy to clean, which makes it an excellent wall covering in areas where the walls get wet.

Electricians and plumbers return after the walls are finished to complete their work. The switches, outlets, lights, and appliances are installed. One master control box is used. All wires in the house are connected to this distribution box. Switches in this box are called *circuit breakers.* The breaker is preset to disconnect the electricity if a problem occurs.

The plumber installs bathroom sinks, tubs, showers, and commodes. Each pipe connection is soldered and checked for leaks. Some appliances, such as a dishwasher, need both the electrician and plumber for installation. The plumber must check to make sure all drains work properly.

Floor coverings are applied after interior walls and ceilings are finished and painted. Hardwood flooring is installed with a special nail and nailer. This flooring has tongues and grooves which prevent cracks between boards. The nails are driven into the edge at an angle. The nail heads do not show as a result. The flooring must be sanded smooth with a belt sander. Then a clear protective coating is applied.

Carpet is used in many homes. It is available in numerous colors and patterns. Nylon, polyester, and vinyl are examples of carpeting materials. Some carpets such as vinyl have to be fixed with glue. These materials are used in the kitchen area. Vinyls are tough and easy to clean. Polyester and nylon carpeting are installed over padding. This gives the carpet added softness. This carpet is held in place by stretching it over tacking strips. These

Houses at different stages of construction.

strips are nailed around the area to be carpeted and then the carpet is stretched into place.

Trimming around doors, windows, baseboard, ceiling, stairs, and cabinets will conclude the carpenters' work.

Painting and finishing can be done to the interior and exterior. Windows, doors, siding, etc., must all have a protective coating.

Windows and doors are sealed on the outside with a weather proofing material such as caulking. The thick paste is squeezed into the cracks making an airtight seal.

Landscaping

Landscaping a new house is a very important last job. Seeding or sodding the lawn and planting shrubs and trees give the house a pleasant look. A driveway and sidewalk are also added to complete the final appearance. The soil around and under the house is treated for termites in certain areas of the country.

A final inspection of electrical wiring and plumbing completes the legal responsibility of the county or city. Utilities are connected, and the house is inspected by the builder. Any items not completed will be scheduled for completion. Upon completion, the house can be transferred to the new owners. This is a legal process. The owner usually has a lawyer draft the necessary legal forms. The lawyer also records the new deed at the clerk's office. After this, the new owners can move in any time.

The money needed to build a house often is borrowed. A loan company or bank loans the money. A fee, a percentage of the borrowed amount, is charged for this loan. The loan is called a *mortgage*. The fee is called *interest*. The mortgage and interest is paid in monthly payments. The length of payment is usually twenty to thirty years, depending on the amount borrowed and the amount of each payment.

Working in Construction

Education

The demand for trained personnel is increasing. The employees in construction may be skilled or semiskilled. The skilled workers are called journeymen in their field such as carpentry, painting, or electricity. A semiskilled worker does work such as carrying bricks or block, pulling wire, and cutting pipe.

A high school diploma is the minimum requirement for a full-time worker. A trade school education in a construction area will start the worker at much higher wage and skill level. The level depends on the training received. A high school graduate could start laying brick at $8.00 to $12.00 an hour. More training and experience is needed to become a skilled journeyman bricklayer.

The professional jobs in construction require a college degree and/or state licensing. Architects, surveyors, and engineers must study their profession many years. These jobs require a great deal of technical knowledge. Designing and building an office building is a big responsibility, requiring highly trained professionals.

Construction contractors are business men and women. They must manage labor crews, buy materials and equipment, and plan and direct construction projects. Construction superintendents are needed to manage construction projects in the field.

Training

Unskilled workers are usually shown how to perform a job. A simple job like digging a ditch requires some training. This training is called *on-the-job training* and usually involves a short period of time.

Skilled training programs are found in some areas. They are called *apprentice training programs.* These programs are offered through the labor unions. The employee applies for the positions and then receives training on the job under another journeyman. Night classes are given in theory and technical information.

Sanding and finishing are just two of the skills needed by a carpenter.

Trade schools train workers in several fields. Brick laying, carpentry, electricity, and plumbing are examples of two-year courses. These schools often require their students to complete high school before entering. It also is possible to learn a trade in a high school level vocational school.

Advancement
Advancements and raises are given with experience. Carpenters and other tradesmen can progress to leaders in their field. These positions would pay top wages. Some workers form a business by subcontracting their labor and materials. These services are needed by contractors.

Apprenticeship employees start at a higher wage than unskilled employees. Increases in pay are regular. At the completion of the program, the trainee becomes a journeyman. A journeyman's wage will be union scale. This means the employee can expect the same wage as all other journeymen in that field of work.

Job Entry
To get a job in construction the following steps are recommended:
1. Submit a letter of application or fill out an application with a contractor
2. Submit a résumé of your background and experience
3. Submit letters of reference from former employers
4. Interview with personnel manager
5. Follow-up.

Construction means not only the building of houses, but also of dams, bridges, tunnels, and airports.

Letters of application may be mailed to the construction companies. These letters are a way of learning about possible positions or job openings.

Occupational Opportunities

The outlook for construction workers is good. However, construction work is seasonal. This means some workers do not work during the cold winter months. A bricklayer cannot lay brick in freezing temperatures. The economy also may affect construction workers. When the economy is slow, construction jobs are hard to find. The number of building permits issued in a month tells the story. Large numbers mean there are plenty of jobs. Small numbers mean job lay-offs. The term seasonal really describes the employment picture in construction. Workers must be willing to move with the work. Some workers travel weekly to the site and return home on weekends.

Wages in construction are among the highest of all workers. Each of the trade unions—plumbers, electricians, and so forth—bargain with management for top pay. The construction unions are trend setters. This means other unions compare their wages to construction unions. When there is little work, unskilled workers in construction are the first workers to lose their jobs, or be "laid off." These jobs are usually filled by local workers. Unskilled construction jobs are over once the job is completed.

Suggestions for Further Study

Personal Involvement

Many activities will help you study construction. Use these examples as a guide:

1. Take a field trip to a new housing development. There you can study the stages of house construction first-hand. Note: Care should be taken. A construction site is very dangerous. Hard hats are required on all construction sites.
2. Select a unit to study in detail. Studying the land development could involve building a topographical map. The activity manual will explain how.
3. A school construction project could be a class project. Often some small construction job at school is needed. A ticket booth, storage shed, or shelter are examples. The activity manual includes plans for building a barn tool shed. This building can be built in school and taken to a site and erected.
4. Conduct a study of the construction industry. Let each class member select a building project. Some examples are: bridges of all types, houses of all types, civic centers, airports, factories, schools, government office buildings, apartments, or roads. Each class member will study one form of building. Encyclopedias may help you to find information.
5. Clip newspaper articles about construction projects in your neighborhood. Then post them on a bulletin board.

From this information a student could construct a model. The model could be built on a scale of two centimeters equals one meter (ratio 1:50). Model building materials can be found, made, or purchased. The finished products could be arranged into a large community.

Self-Assessment

1. Who were the first builders of arches, roads, and bridges?
2. What effect did construction technology have on transportation?
3. How has technology affected tools used in construction?
4. Describe what an architect does.
5. List some steps in building a house.
6. What is meant by seasonal construction work?

Key Words

Architect	Solar Heat
Asphalt	Management
Pyramid	Foundation
Canal	**Slope**
Bid	Survey
Title	Aluminum
Carpenter	Concrete
Inspector	Insulation

Building and Testing Trusses

Concept: *Triangular Bracing*

Many structures require triangular bracing, called trusses, to make them rigid. Trusses are common in houses, bridges, and towers.

Trusses have been used in construction for centuries. Triangular bracing is believed to have been used by prehistoric humans. The Greeks used designs, many formed in stone, which resemble modern trusses. The Romans used truss designs similar to modern designs in their basilicas. Nearly all architectural styles used truss designs.

Look at various structures around you. Where can you see the principles of truss design being used? In your home or school? How about in a neighboring industry? Look at bridges, towers supporting power lines, transmission towers and arched roofs in churches. Have you seen some of the ideas being presented for future space station designs? (See page viii of the color section.) All of these use many different truss designs to form lightweight but strong structures.

Objective

In this activity you will construct a scale truss and test it for strength. Several of your classmates can build different truss designs. Test each to find which is the strongest.

Procedure

Construction Begin by cutting wood strips for the truss members. You can cut strips of clear pine to near-scale size. Try a scale of $1" = 1'$. This is the scale used by many miniaturists and doll house makers. You may be able to find wood strips cut to scale size in a local hobby shop.

A scale "2 x 4" (which is actually $1\ 1/2" \times 3\ 1/2"$) would be about $11/64" \times 11/32"$. Your scale timbers should be about this size if possible.

Build a simple truss such as the King Post design shown in the sketch. When building trusses, it is important to cut all joints as accurately as possible. Tight, close-fitting joints are important to the strength of trusses. The joints are fastened with connector plates. These are metal plates which are pressed into the wood over each joint.

The connector plates for your scale truss can be made from paper such as index card stock. Cut them to the size and shape shown in the sketch. They can be glued in place.

Testing Support the scale truss between two chairs. Attach a water bucket to the center of the truss. Slowly pour water into the bucket to add weight to the truss. How much weight was needed to break the truss? Be sure to include the weight of the bucket when measuring the weight put on the scale truss.

You will find that it is important to keep your truss in a vertical position. A truss is strongest in this position. It is much weaker when tilted at an angle. How can you hold your truss upright when testing it?

See activity No. 5-4 in the activity manual accompanying this textbook for more information on constructing and testing scale trusses.

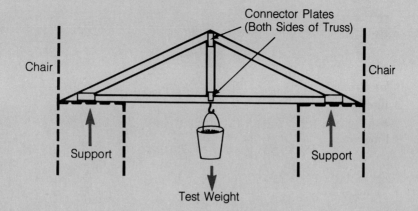

Construction

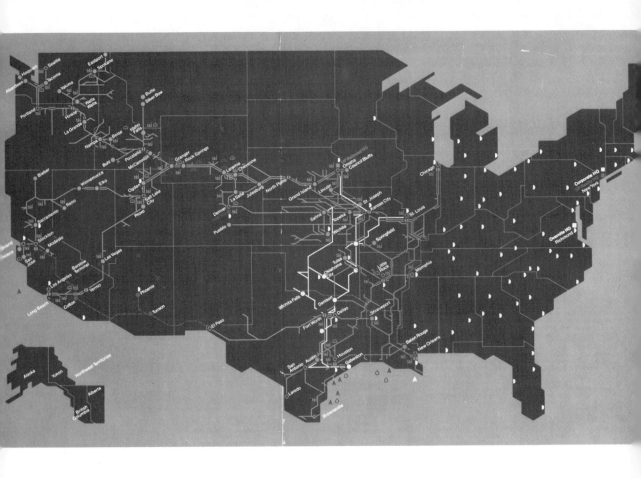

Rail transportation relies heavily on computers for routing decisions.

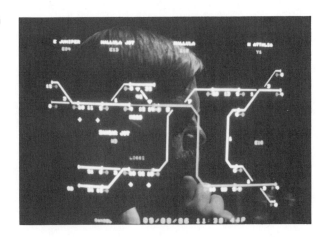

Chapter 6   Transportation

Introduction

Transportation is essential to our way of living. It is also essential for our technological progress. Little that we do today can be done without some form of transportation. Raw materials are transported from their sources to a processing plant and then to factories. Materials, parts, and finished products are transported on conveyer lines inside the factories. Products are transported from the factory to many places where they are used.

Transportation also has a personal importance. We depend on transportation to move us from place to place wherever we want to go. We depend on the automobile, train, bus, ship, and airplane to get us across town, state, or nation or to foreign countries.

Modern transportation is so commonplace today that we often forget it is there. We rarely think about it until it is not there. Nearly everything we use—the food we eat and the clothes we wear—are available to us because of our transportation system.

This is not new, for people have depended on transportation throughout history. Columbus discovered America only after sailing across the Atlantic Ocean. The pioneers of the early 1800s used covered wagons to carry their belongings as they traveled to the western part of our country. The train helped open the West to further settlement and development. Today we depend on transportation for much we consider essential to life.

What would happen if suddenly we had no means of transportation other than ourselves or beasts of burden? What would our life be like? Stop for a minute and think about this. What would your life be like if, suddenly, there were no transportation vehicles powered by engines? What would you have to do without?

As we study transportation technology, it is easier to divide the study into several different parts. These are land, water, air, and space transportation. Each is important in technology. Each is different in many ways and poses different problems. First, to gain a better appreciation of the importance of modern transportation, it will be helpful to know about the development of transportation technology throughout history.

Historical Development Before the Twentieth Century

Land Transportation

The earliest form of transporation was human. When early people wanted to travel or to transport something, they had to use their own muscles. Travel was by walking. Materials were carried on one's back. Then, before the beginning of recorded history, animals were tamed. They were used to carry goods. People learned how to ride the animals. The burden of transporation was taken from the back of the human being to the backs of beasts of burden.

Still later it was discovered that these animals could pull sleds. This made it possible to carry larger loads. The addition of wheels to these sleds was a great advancement in transportation technology. The wheel is considered by many to be the greatest technological development of all time. It is believed to have been developed about 5,000 years ago. The first wheeled vehicles probably rolled on logs. Similar wheeled vehicles can be found in use today in some parts of the earth.

From this simple beginning have come our modern automobiles, trucks, and trains. But development of these vehicles was not fast nor was it simple. Many other technological developments were necessary first. For thousands of years the wooden wheel was used. But it was limited because the wood would break easily. The development of ironworking made it possible to make stronger wheels. Iron bands were put around the wheels. These made the wheels stronger and made them last longer. But, these iron wheels also damaged the roads. This required the development of better road surfaces. Thus, one form of technology, the wheel, brought about the development of another, road building.

All forms of transportation, including land transportation, were slowed by limited power to drive the vehicles. Power was limited to that from beasts of burden. Horses and mules pulled wagons and vehicles that ran on rails like present-day trains.

With the development of the steam engine, all this changed. Speed could be increased. The steam engine produced more power than herds of horses. This marked the beginning of modern

HISTORICAL DEVELOPMENT OF TRANSPORTATION

Time	Development
	Foot travel
	Animals used for transportation
3500 B.C.	Wheel invented
1807	Clermont—Robert Fulton's steamboat launched
1830	First steam-only scheduled passenger train in the U.S.
1861	Otis patents steam-powered elevator
1869	World's first transcontinental railroad completed in U.S.
1893	First American made automobile—Duryea horseless carriage
1900	Orville & Wilbur Wright fly their first glider
1902	"Outboard" boat motor introduced in France and Sweden
1903	Wright Flyer I airplane—first powered, sustained and controlled flight of a heavier than air aircraft
1907	First man-carrying helicopter to rise vertically—Louis Charles Brequet
1907	Harley-Davidson motorcycle company founded
1909	First successful hydrofoil—Enrico Forlanini
1912	Titanic sinks in Atlantic Ocean
1912	First U.S. flying boat—Glenn Curtiss
1914	Panama Canal opens
1914	Electric lights to control traffic installed in Cleveland, Ohio
1915	Tractor-trailer truck invented
1916	Mechanically operated windshield wipers put on automobiles

continued

transportation. As you will see, the steam engine had a similar effect on all forms of transportation, not just land transportation.

Water Transportation

The earliest form of water transportation consisted probably of logs or fallen trees on lakes and rivers and along coastal areas. Later it was discovered that logs could be fastened together to make rafts. These rafts could carry heavier loads. Later logs were hollowed, either with fire or with cutting tools, to form canoes. In some parts of the world, skins of animals were stretched over frames to make boats.

The next big step in water transportation was the use of oars to propel ships. The Egyptians are believed to have had ships as early as 3000 B.C. They used sails and oars for power. The sails could be used to move the ship when it was moving in the direction the wind was blowing. Oars were used at other times. Some of these ships had as many as fifty-two oars and required hundreds of human slaves to pull the oars.

In the fifteenth century ships were built that used sails only. People had learned to sail across and against the wind. Some of these ships could carry as much as 125 tons of cargo.

By the sixteenth and seventeenth centuries, the French were using four- and five-masted ships. Some of these could carry more than 1,000 tons of cargo. The first iron passenger ship was built in 1777 in England. Wooden ships continued to be built until past 1800. In fact, some wooden-hulled troop ships were built for use in World War I.

Transportation developments had far-reaching effects on other forms of progress. Horses made it possible to travel forty-five to sixty kilometers (thirty to forty miles) in a day. On foot it was possible to travel only fifteen to twenty-five kilometers (ten to fifteen miles), and then carrying only a light load. With the horse as a means of transportation, food could be hunted, gathered, and hauled from longer distances. Wheeled vehicles increased these distances even more.

Improved transportation had still other effects on the development of civilization. Small settlements and villages appeared along the transportation routes. People settled and began villages along the navigable rivers, lakes, and major overland routes. These settlements became trade centers for inland exploration and settlement. The villages grew into towns and cities. Along the Eastern seaboard settlements such as New York, Boston, Baltimore, Philadelphia, and Charleston grew into large cities. Ocean

HISTORICAL DEVELOPMENT OF TRANSPORTATION

Time	Development
1917	Sea-going vessel made of concrete
1930	Gas turbine jet engine patented—Frank Whittle
1934	Liquid-fueled rocket successfully fired in Germany—Werner von Braun
1939	First jet airplane flies—German HE-178
1953	Jet airplane flies faster than Mach 1, the speed of sound
1957	Sputnik, first artificial earth satellite—Russia
1959	St. Lawrence Seaway opened
1960	Nuclear powered submarine *Triton* travels around earth in 84 days without surfacing
1961	Alan B. Shepherd makes first sub-orbital flight by an American astronaut
1962	John Glenn—first American to orbit the earth
1969	Neil Armstrong and Buzz Aldrin walk on the moon
1977	Alaska pipeline completed
1981	Train suspended over rails by magnetic levitation runs in Birmingham, England
1986	Space Shuttle explodes 73 seconds after launch—forces new designs in booster system
1986	Voyager airplane—around-the-world, non-stop flight without refueling

shipping provided the transportation for people and materials that made this growth possible. Ocean-going ships from Europe, Africa, and the Far East could unload and load goods for transport to these faraway places. Intercoastal shipping provided communications and trade among the cities and towns along the coast.

Water transportation technology remained at this level of wind-powered ships until the late 1700s. The steam engine changed this.

Air Transporation

Transportation above the earth before the twentieth century was very limited. People have always had the dream of flying. Certainly early human beings watched birds, and wondered how the birds could fly, and why they couldn't.

The first trips by humans in aircraft were in balloons. In 1783 the Montgolfier brothers in France were experimenting with lighter-than-air balloons. On June 5 of that year they launched a balloon made of linen, filled with heated air from burning straw. Later that year the first balloon carrying a person was flown.

In 1785, Blanchard, an Englishman, crossed the English Channel in a hydrogen-filled balloon. He carried one passenger, Dr. T. J. Jeffries, an American. This was possibly the first air passenger flight. Blanchard also made the first balloon flight carrying a person in the United States in 1793. This occurred near Philadelphia.

For many years there had been interest in heavier-than-air craft. (The balloon is considered lighter-than-air). In the late 1800s Dr. Samuel P. Langley experimented with heavier-than-air craft. He tried to use small steam engines for power, but was not successful. However, his work was important to the early development of aircraft.

Some experimenters were flying heavier-than-air gliders. These could be launched from a hilltop and would glide a long distance until they finally landed.

This changed on December 17, 1903, when Orville and Wilbur Wright successfully launched and flew a powered, heavier-than-air craft. This first flight went for a distance of 36 meters (120 feet). The aircraft was powered with a 16-horsepower gasoline engine.

Effect of the Steam Engine

Until the development of the steam engine, transportation depended entirely on human and animal muscles the wind, and the movement of flowing rivers. Speed of transport was limited to the speed of a fast horse. Power was limited to the power that could be generated by a team of draft animals such as horses, mules, or oxen.

During the early 1700s, many people were experimenting with a new source of power, the steam engine. (See Chapter 3 Energy and Power.) In 1763 James Watt made improvements on the steam engine developed earlier by Newcomen. The Watt steam engine was so far superior that Watt is often credited with inventing the steam engine. Watt's engine had great impact on transportation.

The steam engine allowed great advancements to be made in all areas of transportation. This engine could produce great amounts of power. It could produce speeds greater than the horse could run. It could run for long periods of time and didn't have to stop to rest or feed as the horse did.

Land. The steam engine had the greatest impact on land transportation through rail transportation. Some horse-drawn rail carriages were in use. The railways were recognized as a promising mode of transportation. However, it had not been practical because they could not move any faster than any other land transportation. Steam engines changed this. A steam-powered train could move faster and carry heavier loads than earlier ones. These early steam locomotives used wood or, later, coal as fuel.

The *Tom Thumb* was an early steam locomotive.

Water. The steam engine had similar effect on water transportation. Until the steam engine, ships were dependent upon the movement of the winds and currents, or on the muscles of humans and draft animals. This was very limiting. Ocean-going ships could only move when the winds were blowing. River transportation was limited to drifting with the current or being pulled up stream by animals or by human beings rowing a boat.

Steam-powered ships could sail any time, whether or not the wind was blowing. They could move up stream against the river currents. Larger ships could be built because there was more power available to move them and their larger cargoes. Larger loads could be carried.

Air. The steam engine had limited impact on air transportation. At the time the steam engine was being developed, air transportation was limited to lighter-than-air balloons. Balloons could not use steam engines. Early efforts to use steam engines on heavier-than-air craft, (by Langley, for example) generally proved unsuccessful. For one thing, the steam engine was too heavy for the amount of power output. Aircraft required lightweight engines. Heavier-than-air craft would have to wait for some other source of power. This would be the gasoline internal combustion engine.

Effect of Petroleum
In 1859 Sir Francis Drake successfully drilled an oil well in Titusville, Pennsylvania. This marked the beginning of what is probably the greatest change in transportation technology. The oil taken from the earth through oil wells was to become a new energy source. This oil would revolutionize transportation. Oil, and the gasoline refined from it, made the internal combustion engine practical. This engine made low-cost, easily transported engines a reality. The internal combustion engine gradually replaced coal- and wood-fired steam engines in transportation vehicles. The internal combustion engine made the heavier-than-air craft possible. It made the development of the automobile practical.

Modern Transportation by Air and Conveyor Systems

Air/Space
Since the beginning by the Wright Brothers, progress in air transportation has been rapid. In 1909 Bleriot was the first person to cross the English Channel in an airplane. World War I spurred aircraft development. War leaders realized that the airplane could

give their armies great advantage over the enemy. Planes replaced balloons for observation. Airplanes also were used to combat other airplanes. This created a demand for airplanes that were maneuverable. Better airplanes, better engines, and pilots were needed. After World War I many of these war pilots became commercial pilots, flying the mail and freight. Passenger flights in this early era of aviation were usually a novelty. They were done for the thrill of it.

Charles Lindberg was one of many aviators in the early 1920s who dreamed of flying solo across the Atlantic Ocean. On May 21, 1927 he accomplished this. In thirty-two and a half hours, he flew from Garden City, New York to Paris, France. This marked the beginning of a new era in air transportation. As better engines and larger airplanes were developed, long distance freight and passenger flights became commonplace.

With the outbreak of World War II, progress in air transportation increased rapidly. Many technical developments were required before the high-performance aircraft could be built. High octane fuels, pressurized cabins, and larger engines were necessary. These, in turn, made other advancements possible, including large aircraft, bombers and troop carriers and other high-load aircraft. Small, fast, maneuverable fighters were needed. Late in the war the jet airplane was developed in Germany. The war ended before it could see much action or be developed further.

Though many people had experimented with rockets, the age of rocketry also dawned during World War II. The Germans used rockets for long periods of time against England. These were

This Curtiss P-40 was used during World War II.

Transportation

unmanned, but they were the forerunners of our modern spacecraft. Many of the German scientists who developed these rockets were to become the same scientists that developed technology for space program of the United States in the 1950s and 1960s.

Even today space transportation is still experimental. In many ways, it is in a stage of development comparable to the Wright Brothers in aircraft. The space age is said to have begun October 4, 1957 when the Soviet Union launched a satellite, Sputnik I. This was the first artificial satellite to orbit the earth. On April 2, 1961, a Russian cosmonaut, Yuri Gagarin, became the first human being to orbit the earth. This was the first manned space flight. A month later Alan B. Shepard, Jr. made the first space flight for the United States. His flight lasted only fifteen minutes and he did not orbit, just entered space and returned. On February 20, 1962, John H. Glenn, Jr. became the first American to orbit the earth. His flight lasted three revolutions around the earth.

On July 20, 1969, American astronaut Neil A. Armstrong became the first person to set foot on the moon. He was accompanied by Edwin E. Aldrin, Jr. in this historic event. Since this time, progress in space travel has been rapid. Others have explored the moon. A specially designed transportation vehicle was developed to carry the astronauts about the moon's surface. This "moon buggy" allowed the explorers to move faster and to carry more scientific equipment than was possible without it.

As in the development of the airplane, progress in space transportation has been possible only because of many technical developments. New materials that would withstand the high temperature of reentry had to be developed. Powerful rocket engines had to be developed. These engines had to be reliable, so they would not fail while in outer space. Navigation and ways of controlling and navigating the spacecraft had to be developed. Foods were needed that could be carried on long space flights. Objects became weightless in space. Imagine the difficulty one would have drinking a weightless cup of water. This was just one of the problems that required technical answers.

Since these first dramatic space flights, progress in space continues. The *Apollo-Soyuz* flights were the first joint space efforts between the Russians and the Americans. The American *Apollo* and the Russian *Soyuz* craft were launched at about the same time and were able to link in space. The space shuttle is designed as a reusable space craft. The craft can reenter the earth's atmosphere, land on earth, and be used again in later flights. This results in a tremendous saving of money.

Providing passenger service is the major function of airlines. This Boeing 747 is powered by turbofan engines.

Types of Air Transportation

Air transportation can be divided into three major groups: commercial, general aviation, and military. Aircraft are designed primarily for use in one of these groups. Generally, aircraft designed for commercial use are not used in general aviation. General aviation aircraft are usually too small for use by the large commercial airlines. Military aircraft are designed for a special military use and most are not suitable for commercial or general aviation use.

Commercial. Commercial aviation refers mainly to the use of airplanes by the airlines. Airlines provide passenger and freight service to approximately 500 airports in the United States. Most of this service is to the larger cities.

These airplanes provide scheduled service to more than 200 million passengers each year. Passenger service provides approximately 80 percent of the income for the airplanes. The remaining income is received for freight service. Thus, cargo hauling is only a small part of commerical aviation. The cost of transporting freight by airplane is high when compared to surface transportation. Most freight is heavy. This weight adds to the cost per air mile. This makes the cost of carrying freight high. Freight is transported by air only when speed is important. The fast service may make the high cost acceptable.

Transportation

This illustration shows a fuel-efficient aircraft currently under development. Two jet turbine engines will each drive counter-rotating fans, similar to propellors. Increased aerodynamic efficiency and lighter construction will allow for lower operating costs.

General Aviation. All nonmilitary aviation except that of the commercial air carriers is called general aviation. General aviation includes flying by businesses (72 percent), personal transportation and proficiency (practice or training) flying (23 percent), and sport flying (5 percent).

Small, general aviation aircraft are used to provide air taxi, rental, and small commuter operations. People can save time by chartering or renting airplanes to fly them where they want to go. It is often an advantage to be able to travel by air when one wants, rather than having to wait for a scheduled airline flight. It can be faster because there is less time lost getting to a large commercial airport, which may be some distance away. It also can be faster because the flight can be direct, without the stops scheduled by the airlines.

Many small companies provide air taxi and air freight service to all parts of the country. Many of these connect with scheduled airlines at the larger airports. Air taxi service is moving rapidly closer to each person's home. More small airports now are being built to provide this local service. There are now more than 16,000 airports in the United States. Most of these are small and offer local service. 11,000 of these are privately owned airports.

Personal Transportation and Proficiency. Some people own their own airplane or rent one to use. Flying clubs offer pilots the chance to use the club's airplanes at less cost than to nonclub members. Private pilots use an airplane almost as they would use the family car for family or business trips. A vacation trip by personal airplane would take less time, use less fuel, and leave more of the vacation time for fun. Less of the vacation time would be spent traveling.

Pilots must maintain their flying skills. Federal regulations require that a pilot fly a minimum number of hours each year in order to keep the pilot's license. This, along with flying for instruction or getting an advanced license, is called *proficiency flying.*

Business. Businesses use aircraft to save time and money transporting executives, salespersons, repairers, and customers. A business executive can travel much faster visiting the company factories and warehouses around the country. Company airplanes transport passengers to the larger airports where they can connect with scheduled airline flights to the big cities for long distance flights.

Off-shore oil rigs are supplied in part by helicopters. This is an example of special-purpose flying.

Special-purpose flying includes the transportation of large loads. Here a two-engine helicopter lifts a medium-sized airplane.

Broken machinery can close a plant for a long time. Money is lost while repairmen wait for parts to arrive. The parts can be transported by airplane in much less time than is possible by surface transportation. Repairs can be made and the factory put back into operation sooner by using air freight to ship parts.

Special-Purpose Flying. There are numerous reasons for flying that do not fall under the above categories. These are called special-purpose flying. Air ambulances transport critically ill persons to hospitals for treatment. Helicopters are being used to carry victims of traffic accidents to hospitals. Agriculture has benefited greatly from the use of airplanes. Crops can be seeded, fertilized, treated with insecticides, and checked for disease from the air. Fields of grain covering hundreds of acres can be sprayed in a short time by crop dusters.

Sport. Sport flying is only a small part of the total general aviation flying, but it is an area important to many people. Sport flying is flying just for the fun of it, to get away from the day's pressures and to enjoy the freedom flying gives. About 5 percent of general aviation is for sport.

Military. Military aircraft are designed to serve one or only a few special purposes. In some cases general aviation or commercial aircraft are purchased by the military and adapted to suit the

Transportation

military needs. Passenger airplanes are used sometimes to transport troops, but most military airplanes are specially designed for military use.

Fighter planes are designed to be fast and maneuverable to intercept other fighter planes. Bombers are made so they can fly long distances and carry heavy bomb loads. Transport planes are designed to carry heavy loads of different items. Some can be used to carry cargo or have seats installed to carry troops. Some can be made to carry wounded soldiers to hospitals far from the combat zones. The Navy uses specially designed fighters and light bombers that can take off and land on aircraft carriers.

Conveyors

When we think of transportation, we usually think only of cars, trucks, trains, ships, and airplanes. But transportation is more than this. Transportation includes the movement of materials through pipelines and other conveyor systems such as elevators and escalators. These are a part of transportation technology that we can not overlook in our study of technology.

Our modern way of life depends on these important forms of transportation. The water we drink is transported in pipelines. The gas used to heat homes and industries is transported in pipelines. Fuel oil and gasoline also may be transported over long distances by pipelines to storage tanks in the areas where they are to be used.

Conveyors are important ways of transporting materials in factories and the parts to an assembly line. The assembly line itself is a conveyor. After assembly, the finished product is carried

An overhead conveyor moves parts through a manufacturing plant.

Crude oil flows through pipelines and valves from the oil field to the refinery. Finished products flow from the refinery to consumers.

to the packaging department and then to shipping. Often the packaged product is moved into trucks or train cars on some type of conveyor.

Some conveyor systems are being built to move people. These are often called *moving way* or *moving sidewalk* systems. These are similar to conveyor belts but are made large enough and strong enough to carry people. These are being used in airport terminals to move people to the waiting airplanes.

Pipelines. Pipelines are used to transport liquids and gases. They often are run below the surface of the earth or under water. They often are, built above the earth or held above the surface by cables, posts or pillars.

Pipelines are systems. They are made of several parts: the network of pipes, the terminals for storage, pumping operations, control and distribution operations, and storage tanks. The network of pipes is much like the network of transmission lines you learned about in Chapter 3, Energy and Power.

Years ago some pipe was made from logs with holes drilled through the center. Modern pipes, 7.5 centimeters (3 inches) and larger, are made of cast iron, reinforced concrete, or an asbestos-cement mixture. Smaller pipes are often made of iron, plastic, copper, or steel. Cast iron, glazed tile, and concrete are used for drain and sewer pipe.

Pipelines can be used to transport solid materials too. As early as 1913 coal was mixed with water and moved through a pipe in England. From a wharf on the Thames River, the coal was moved

594 meters (660 yards) to the Hammersmith Generating Station. Some of the pieces of coal in this pipeline were as large as 12.5 centimeters (five inches) across.

Since then it has been found that smaller sizes of material can be moved more easily in pipelines. Coal can be moved in a pipeline efficiently if it is first crushed into pieces no larger than 0.3 centimeters (1/8 inch). It then is mixed with water and pumped through a pipeline. At the end of the pipeline, the water is removed and the coal dried. The coal then can be used for fuel.

Not all materials can be transported by pipeline. Some are too heavy. Others cannot be mixed with water without ruining them. In many cases the cost of the process is too high. The added process of grinding the material before transport and then drying it after transport often makes it uneconomical to transport by pipeline. The cost of building the pipeline also adds to the cost of transporting the material. A pipeline must be used for many years before its cost can be recovered.

Conveyor Belts. Conveyor systems are used for the mechanical handling of many materials, bulk and packaged. *Bulk handling* is the handling of materials in bulk or loose form such as coal, grain, ores, sand, or gravel. *Package handling* is the handling of materials that have been put into some type of container.

Conveyor belts usually are used to transport bulk materials. The most commonly used belts are made of canvas, rubber, or some kind of plastic material. The conveyor belt is actually a system; it is made up of several parts.

Simplified Conveyor Belt System

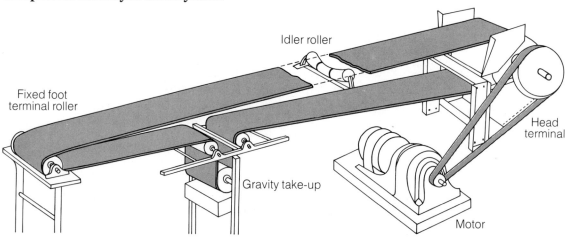

The belt is the main part. It can be made of various materials, depending on the way it is to be used. Belts used to carry small parts or small bulk materials such as sand are made of finely woven materials such as canvas, rubber, or one of the plastics. The finely woven belt is needed to keep the small parts or particles from falling through the belt.

Woven wire belts are used where the belt must run through extreme temperatures such as those in a bakery. In the bakeries the loaves of bread are put into pans. Woven wire belts carry the pans of bread dough through the oven, moving at a set speed so that the bread is baked when it comes from the oven. Woven wire belts are also used to transport materials through spray booths, or to allow liquids to drain from objects. Where the circulation of air is important, the woven wire belt again is used.

Large rollers, called terminal rollers or pulleys, are used at the end of a conveyor belt. These rollers are connected to electric motors which provide the power to the conveyor belt.

Rollers are needed in a conveyor system to support the belt and keep it running straight. These rollers are not powered and, thus, are called idler rollers. They *idle* or are unpowered. Some idler rollers shape the belt into a trough shape so that loose materials will not fall off the edges. This is usually done when transporting materials such as sand, gravel, ores or coal.

It is necessary to keep the conveyor belt tight on the rollers. This often is done by a *gravity take-up*. This device is placed in the lower or return part of the belt. The weight on the take-up pulls the belt tight and keeps an even tension on it.

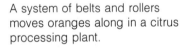

A system of belts and rollers moves oranges along in a citrus processing plant.

Some conveyor belt systems are made for special uses. Slat conveyors are used to transport packages or large items. Wooden or metal slats are used instead of the woven belt. These slats are fastened to a chain which runs around sprockets. Usually the chain is supported by some type of guide or trough. Often rails are placed along the sides of the slat conveyor to prevent the materials from falling off the sides.

Roller conveyors are used for heavy and hard applications such as those in steel mills, used to move large pieces of metal or those in foundaries, used to move large castings. Roller conveyors are also good when it is necessary to stop materials on the conveyor. Roller conveyors can be made in short sections so they can be taken apart and moved. This makes them useful in loading and unloading trucks and train cars.

Some roller conveyors are powered by electric motors, making it possible to move materials up an incline or horizontally.

Often it is necessary to move materials vertically from one floor of a plant to another or to the top of a machine. Bulk materials are often carried in vertical elevators or *conveyor-elevators.* Buckets are suspended between two endless chains running on rollers. The buckets carry the bulk material vertically and then can be tipped over so the material is dumped into a hopper or onto a belt, which carries it horizontally.

Screw conveyors are used to transport bulk materials of fine and medium size. The screw conveyor has a long screw mounted in a trough or pipe. As the screw turns, the material is pushed forward by the sides of the feed screw. Screw conveyors are used where accurate amounts of material must be moved. The screw conveyor also helps break up materials that tend to pack or form lumps.

Escalators. An escalator is actually a conveyor system. It is designed so it forms a moving stairway for people. It makes it much easier for people to move from one floor to another in a building. If the escalator should stop running, the steps still can be used as a stairway.

Moving Way. The moving way is a conveyor belt that has been designed to transport people. Some modern airports use the moving way to move people between ticket counters or baggage areas and the airplanes. Often it is a long distance from the ticket counter to the waiting room or the boarding gate. Many people find this long walk tiring, especially if they have a lot of baggage.

This moving way transports people around the San Diego Zoo.

Components of Transportation

By providing a moving walkway, the people only have to step on and off at any point. In this way, the moving way provides smooth, continuous movement of the traffic.

So far you have seen various ways of transporting materials, you have studied some of the history of various forms of transportation, and you have seen how land, sea, and air/space transportation developed.

Now you will learn about the various components that make up a transportation system. *Vehicles* are the carrying part of the system. The vehicle is the automobile, the airplane, or the ship. *Power plants* are the source of power that move the vehicles. *Controls* are needed to keep the vehicle traveling in the proper direction. The *way* refers to the route the vehicle travels—the roadway, railway, air lanes, and shipping lanes.

Vehicles

Transportation vehicles are designed to serve special needs. Automobiles, for example, are designed to transport people. The comfort of the people is important. The seats are designed for comfort. The controls are placed so the driver can see and reach them easily. Trucks are designed to haul freight. Only the cab of the truck must be designed for human comfort. The trucks also are made larger than cars. They must haul large loads so the trucking company can make a profit.

Ocean liners are designed to transport people. They are made so the passengers can be comfortable and enjoy their trip. Cargo ships are designed to carry as much cargo as possible. Only a small part of the cargo ship must be designed for people. The crew that operates the ship lives in quarters designed for them. The rest of the ship is designed only for cargo.

Transportation vehicles also are designed for the way they must operate. Land vehicles are different from water vehicles. Air and space vehicles are different from land vehicles. Air vehicles must operate in the earth's atmosphere. Water vehicles must operate on or under water.

A land vehicle cannot operate on water unless it is designed for both media. Then it is called an *amphibian*. Without gravity a car, truck, or train or ship could not operate. It would float off into space. But an airplane must overcome gravity. The airplane wing is designed to provide the lift needed to overcome gravity.

Cargo ships have many different designs depending on their intended use.

Transportation 197

Transportation vehicles must also be designed for the places they are to be used. Trains, ships, and large trucks cannot operate in the inner city. But they can carry passengers and cargo to central terminals. Smaller trucks, buses, and cars can carry the people and goods to the inner city. Large passenger airplanes fly between the large terminals. Buses, taxis, and even helicopters carry the people and cargo into the city itself.

Water/Ocean-Going Vehicles. Most ocean-going ships in use today are cargo ships. Passenger liners are still in use but are only a small part of the total ocean-going transportation. Airplanes have taken most of the ocean passenger trade. Most modern jet airplanes can travel from New York to London in seven hours. Some can travel that distance in about three hours. It takes a passenger ship several days. Most people today want the fastest transportation possible.

For this reason, most of the discussion in this chapter will be about cargo transportation. You may want to do your own study of passenger liners. Those that are still in use today are like small floating hotels. They have similar services. They have restaurants that prepare and serve meals. Shops and stores sell their goods to crew and passengers. Recreation includes swimming pools and games such as volleyball and badminton.

The *passenger liners* operate on regularly scheduled passenger service. They carry as many as 2,000 passengers. They also carry cargo. The regular schedule is important. They leave port at a scheduled time and arrive in the next port at a scheduled time. Just like airplanes or buses carrying passengers, the liners keep their schedule whether they have a full load of passengers or not. Ocean liners usually travel between places where a large number of passengers are available.

Cargo liners are cargo ships that also carry passengers. They operate on a regular schedule. Cargo ships can carry as many as twelve passengers without meeting the very strict safety regulations for passenger liners. For this reason cargo liners seldom carry more than twelve passengers.

Tramps are freighters or cargo ships that do not operate on a regular schedule. The route they travel is determined by the cargo they find. Tramps usually haul bulk cargo such as coal, grain, lumber, sugar, or ores.

Cargo ships can be classified as general cargo, tankers, dry-bulk carriers, and multipurpose ships. There are also many types of special-purpose ships. Specialized cargo ships are

A tugboat guides a passenger liner out of the harbor.

designed to carry one certain type of cargo. Tankers are specialized ships. They can carry only liquids such as oil. Ore carriers are specialized cargo ships, designed to carry the very heavy iron ores. These ships must be built stronger than other ships. Most of the specialized cargo ships are tankers. Some carry crude oil to the refineries. Some carry products from the refineries to distribution points.

General Cargo. Many tramp ships are general cargo ships. They are designed to carry a variety of cargoes. Usually the cargo is made up of many different items. Cotton may be shipped in bales. Coffee may be in bags or boxes. Automobiles may be shipped uncrated. Machine parts may be crated. The general cargo ships may also be carrying items such as fruit, sugar, cement, flour, fertilizers, or machinery.

The general cargo ship must be designed so that these materials can be stored with little waste of space. Usually the general cargo ship has cranes on deck to load and unload the ship. Unless some special cargo is on board such as heavy machinery, no cranes are needed on the dockside.

Specialized Ships. Specialized ships are designed to carry only one type of cargo. For example, container ships are designed to hold large containers of cargo. The containers are huge boxes of the same size and shape that hold the cargo. A shipper can load the containers and seal them. They then are carried on trucks to the docks. Loading the ships can be done quickly. Handling of cargo is reduced, thus reducing labor and cost. Containers also help prevent theft and damage of the cargo.

Crude oil is the cargo of this huge tanker.

A *LASH* (Lighter-Aboard-Ship) is a specialized ship. This type of ship is designed to carry small barges. The barges can be loaded in shallow water ports. They are towed to the LASH ship in the deep water port and loaded. The ship carries the barges to a deep water port near their destination. The barges then are unloaded from the LASH and towed along the inland waterways to the places where the cargo is needed. This provides a door-to-door service for the cargo. The LASH ships also can carry containers.

Tankers are also specialized ships. They are designed to carry only liquids. Most tankers are owned by the large oil companies. The ships operate on regular schedules. The tanker is divided into many smaller tanks. Each tank is watertight. Each tank must also be ventilated to allow gases and fumes to escape. Pipelines run from each tank to pumprooms. Here pumps move the liquid between tanks or load or unload the liquid at dockside.

Tankers do not need a lot of loading or unloading equipment. The docks have pumps which can load or unload the liquid

Iron ore tumbles from the loading dock into the holds of an ore-carrying tanker.

quickly. This allows for quick turnaround (time spent in port waiting to unload, reload, and continue the trip). Tankers carrying oil often carry cargo in only one direction on their trip. They carry oil from the oilfields but there is no liquid to be carried back to the oilfields. They must return loaded with ballast. This ballast is usually water. The ballast is needed to weigh the ship into the water far enough to make it move properly through the water. An empty ship would bob around in the ocean like a cork. Heavy seas could damage an empty tanker.

In recent years ships have been designed to carry several types of cargo. These are called *O.B.O.* (ore/bulk/oil) ships. These ships can carry ores into a port, unload, clean the holds, and take on a load of oil. This oil is carried to another port and unloaded. The holds are cleaned and can be filled with bulk cargo such as grain, salt, wood chips, or other loose cargo. Cargo is not mixed in these ships. Oil and grain are not carried in the ship at the same time.

Some multipurpose ships are designed to carry mixed cargoes. Some have one hold that is refrigerated to carry frozen foods. Another hold is designed to carry automobiles. Another may be designed to hold liquid cargo or general cargo. These ships must be equipped with cargo handling equipment such as cranes, pumps, or ramps for roll-on/roll-off vehicles.

Barges are flat-bottomed boats pushed or pulled by tugboats. Some barges are powered by their own engines. On inland waterways such as the Great Lakes or the Mississippi River, barges are used to carry large quantities of cargo. Bulk cargo such as coal, ore, lumber, logs, and steel is hauled in the barges. Often several barges are connected end-to-end in a unit called a *tow* and pushed or pulled by tugboats.

Shipbuilding. As you study this section on shipbuilding, keep in mind that nearly all commercial ships are built in the same way. They may look different. They may be different sizes and shapes. They may carry different cargo. But they are built in much the same way.

The differences in shape and size happen because ships are designed for special purposes. Each ship is designed individually. Most ships are designed to travel over a certain route or to carry a certain type of cargo.

Shipbuilding is quite different from what it was just a few years ago. The use of steel in ships has brought about this change.

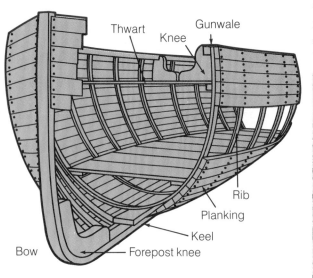

left Rib and keel construction of a wooden boat. *right* Prefabricated sections are welded into place during the construction of a double-bottom steel ship.

Early ships were made entirely of wood. The earliest ships even used wooden pegs to fasten the parts together. Wooden ships were built by first laying a keel. This was a long wooden beam that ran the length of the ship, from bow to stern. Ribs were added to give the ship its shape. This was the "skeleton" of the ship. Planking was added to the ribs. This "skin" was the sides of the ship and made it watertight. Decks were added, and the ship neared completion.

Some wooden ships had metal plates fastened to the wood sides. These plates helped protect warships from cannon balls and musket fire. Gradually the ships were made entirely of metal. Metal keel and ribs were used. Metal plates were welded to this skeleton.

Today all commercial ships are built entirely of steel. A keel is not used. Instead, a double-bottom assembly is used. The double bottom is made like two bottoms. These are separated by bulkheads. The entire bottom of the ship gives more strength than a keel. The double bottom can be used to carry liquid cargo or fuel for the engines in the ship.

Modern ships are built in prefabricated sections. Each section is made in a shop in the shipyard. Often the sections have pipes and wires installed while in the shops. As the sections are completed, they are carried to the assembly area by huge cranes. The sections are welded together to form the hull of the ship.

The superstructure is added to the hull: the bridge, crew quarters, and essential equipment. The completed hull is now ready to be launched.

Most smaller ships are built on a shipway. The ship slides down this shipway into the water when it is launched. Launching is done stern first. If a shipyard is on a narrow river, the ship may be launched sideways. The larger ships such as aircraft carriers are built in dry docks. At launching, the drydock is flooded, and the ship is floated out.

After launching, the ship is just an empty hull. Tugs pull the hull to an outfitting pier. Here the thousands of small furnishings are added. Outfitting may take as long as a year to complete.

After outfitting, the ship must go through a series of sea trials where all equipment is tested. Speed is tested over a measured course. The amount of fuel used by the engines is checked. Tankers are filled with sea water to test pumps, tanks, and the ship's handling with a full load.

If the sea trials show needed changes or repairs, these are made. Then the ship is ready to be turned over to the new owners. The ship is now put into service with a full load.

Land-Transportation Vehicles. Land-transportation vehicles can be divided into three main types: automobiles, trucks and buses, and trains. There are also many special purpose vehicles. Automobiles are used mainly to transport people. Usually only one or a small number of people are transported in these vehicles. Trucks transport large amounts of cargo. Buses transport large numbers of people. Trains transport people and cargo. Today most trains are freight trains. The passenger business of trains is small. Special-purpose vehicles include tractors and other farm equipment, construction equipment, mining vehicles, and school buses.

Automobiles. Automobiles have become one of the most important means of land transporation. In the United States the automobile has become a way of life for most people. People depend on it. For many people an automobile is a necessity, but this has not always been so.

Before the automobile was developed, people walked or bicycled or used animals for transporation. Few people traveled very far from their homes. They seldom traveled outside their community. People worked near their homes. Workers often lived in the building in which they worked. Often companies built houses near the factories so the workers could live nearby. These people did not need transportation vehicles. All this has changed. Today, many people live many kilometers from their work. They commute to work each morning and home again at night. In the larger cities, commuters may travel as far as eighty kilometers (48 miles) each way. The development of the automobile and good highways has made this possible. A person can now travel a distance in one or two hours which used to take a day to travel. A person can walk comfortably about five kilometers (3 miles) per hour. On horseback, a person can travel about 16 kilometers (9.6 miles) per hour. A person driving an automobile can travel over 88 kilometers (55 miles) per hour.

As the automobile, train, and bus were developed, people were able to travel more. They could travel greater distances in less time. Workers could live farther from their work. People came to depend on vehicles to transport them.

The development of the automobile is an interesting story that is an example of the development of many new products. Many new materials and processes of working them had to be developed before the modern automobile was possible.

You may wish to study the development of the automobile more deeply. Ask your teacher for help. Check in your school or community library. Many books have been written about the development of the automobile. For our study we will concentrate on the major components and systems in the automobile.

New automobiles must be transported to dealers. These special-purpose trucks are loaded in large yards near the factory.

Trucks, buses, and all other road vehicles have developed from the automobile, the passenger car. The first road vehicles were designed to be passenger cars. As these developed, other vehicles were designed. Trucks were developed to carry cargo. Buses were designed to carry large numbers of people. Highway construction equipment was needed to build highways for the autombiles to run on. This equipment was patterned after the automobile. Other than trains, most land transportation vehicles are patterned after the automobile. Even the moon buggy looks much like a passenger auto used on earth.

Keep in mind that all land vehicles are basically the same. Cars, trucks, and buses all have similar parts. All have a frame, body, and suspension system. The truck and the bus are large, heavy, and less streamlined than the automobile. But the truck and bus are made up of the frame, body, and suspension system, just as the automobile. Each vehicle is designed specially to serve a special need. This special need makes each vehicle look different. As we study land transportation vehicles, our study will concentrate on the automobile. Keep in mind that other land vehicles are similar.

As has been done many times in this book, a complex topic will be broken into smaller parts for study. Each part can be studied separately. This makes it easier to understand the bigger topic.

The automobile is made up of the body and frame, the wheels and tires, and the suspension system. The suspension system connects the axles and wheels to the frame and body.

The body is designed very carefully. It is streamlined for beauty and to reduce wind resistance. Many people will buy an automobile because of its design. They like the look of the body of the automobile. The design of the body changes from year to year. Compare one of the early automobiles made by Henry Ford with one of the modern autos. There is considerable difference in the styling.

The frame of an automobile is similar to a skeleton. The rest of the vehicle is built around the frame. When an auto is built, the frame is assembled first. The power plant, drive train, and wheels then are added. This is the chassis. All other parts are then added to the chassis.

The suspension system in the automobile helps give a smooth ride. The springs and shock absorbers (shocks) help smooth out bumps in the road. The shocks help retard the bounce of the springs.

A car body in an assembly plant is conveyed toward the area where it will join the chassis.

The automobile is made of several subsystems. The power plant and drive train make a power system. The steering system and braking system are subsystems of the control system of the auto. The auto engine is composed of several subsystems. The cooling system keeps the engine cool. The lubrication system keeps oil on the moving parts to reduce wear. The electrical system provides electricity to the engine and for lights and various powered accessories.

The assembly of the automobile is among the best examples of assembly-line production. The assembly line was started by Henry Ford. (See Chapter 4, Manufacturing.) Unlike other transportation vehicles, the automobile is mass produced. Many are made that are identical. This is not true in other transportation vehicles. Seldom are two ships made alike. Many airplanes are alike, but not as many airplanes are built as automobiles.

Trains. Trains are not as new as many people think. As early as 1465 wagons running on wooden tracks were used in German mines. In 1810 steam engine trains were used in English mines. Until this time horses had been used to pull the coal out of the mines. By 1848 there were 8,000 kilometers (4,800 miles) of railroad tracks in England.

Since these early days the train has become an important part of our transportation system. At one time in the United States trains were the major form of long-distance transportation. The automobile and airplane have replaced the train for this purpose. The airplane offers faster, more comfortable transportation. The automobile offers convenience. People can drive door-to-door in the automobile. They do not need to find transportation to and from the train or air terminal.

But, in these times of crowded highways and fuel shortages, the train may come again as a major form of transportation. New and modern trains may resume some of the job of transporting people and materials.

The train is made of two parts—the locomotive and the cars. The locomotive is the power plant in most trains. It will be mentioned only briefly here because it will be discussed in more detail in this chapter in the section on power plants.

The train cars make up most of the train vehicle. Unlike other transportation vehicles, the train is made up of many similar parts, the cars. When a train is put together, it usually is made of only passenger or only freight cars. The passenger train usually carries only passengers. Often there is a baggage car to carry the baggage carried by the passengers. Some passenger trains include cars that carry automobiles. Passengers can drive to the train terminal and put their auto on the train. When they arrive at their destination, their automobile also is there to be used for local transportation. In this way people from New York can travel to Florida for a vacation and have their personal automobile to use while there.

Sleeping cars are designed for overnight travel. In some cars the seats can be made into beds. In others private rooms are used by the passengers. The seats in the rooms can be made into beds. Beds also can be pulled down out of the ceiling of the room.

Passenger cars provide only seats for the passengers. Some of these have a large glass dome in the top of the car. The seats are raised so that the passengers can see the passing scenery through the dome. Often a passenger train will include cars equipped with lounge chairs or sofas. Some cars have tables for games or other forms of entertainment during the trip. Long-distance passenger trains also have *dining cars*. These are like restaurants on wheels with tables, chairs, and waiters serving the food.

Freight trains are made only of cars for hauling freight. They are joined in a huge classification yard. Here each car is pushed onto a hill in the yard called a hump. From here the car rolls down tracks into the yard. The yardmaster switches it onto different tracks until it rolls to the proper train. It is then coupled to the train. The train is made of the proper cars, often in a certain order. A caboose then is coupled onto the back of the train, which can pull away, headed for many miles of travel and many stops.

top left Hopper car. *top right* Passenger coach.

Some freight cars are general-purpose vehicles. They can be used to haul a number of items. The boxcar is the general cargo vehicle of the train. It is completely enclosed. It is sealed from the weather and the doors can be locked. In some cases the cargo is packaged in boxes, bags, or crates. Sometimes items such as tin cans, wood chips, or plywood are loaded directly into the boxcar. The *boxcar* is used for materials that must be protected from the weather.

Freight that needs no protection from the weather can be carried in one of the open freight cars. Bulk materials like coal, ores, or crushed stone are carried in open *gondola* or *hopper cars*. These have doors in the bottom which can be opened to dump the cargo quickly. Often lumber or pulpwood is carried in gondola cars.

Flat cars are used to carry large machinery or other large freight. Today many flat cars are being used to carry "piggy-back" trucks. These are truck trailers that are driven onto the flat car. The trailer is then carried to a distant terminal. There, a truck tractor is hooked to the trailer, and its contents are delivered to the destination. In this way the trains provide containerized service. A manufacturer can load a "piggy-back" trailer in the factory and seal it closed. The trailer is pulled by a trucker to the main

Basket car.

208 Exploring Technology

terminal. The manufacturer also may have a train siding at the factory. The trailer is loaded on the train flat car and transported to the terminal or siding nearest the destination point. The train is then unloaded and pulled to the customer by another trucker. The cargo arrives in its original container, still sealed and protected from damage and theft. It has been handled a minimum of times.

Many train cars are special-purpose vehicles. They are designed to carry only one kind of cargo. *Tankers* are designed to carry certain types of liquids such as milk, acids, oil, gasoline, or fuel oil. *Refrigerator cars* are made to carry perishable foods. Some have compartments that hold ice to keep the contents cool, while others have refrigeration units similar to home refrigerators. Cars carrying frozen foods have refrigeration units to keep the foods frozen.

Stock cars for animals are similar to boxcars. The sides are left partially open so the animals inside can get proper ventilation. Often these are two or three level cars, so more animals can be carried in a single car.

In recent years the *auto rack car* has been used to transport new automobiles from the factory to the auto dealer. Often these are two or three level cars. Some can carry as many as fifteen automobiles. It is not unusual to see as many as twenty to thirty of the auto rack cars in a train. By using this type of car, a large number of automobiles can be carried a long distance in a short time.

Air Vehicles. In the short history of aviation, airplanes have changed dramatically. The airplane the Wright Brothers flew was small, flimsy and made of wood and cloth. The 16-horsepower engine was homemade. The pilot lay on the center of the wing.

From this rather simple beginning have come modern supersonic aircraft made of aluminum and exotic metals, powered by jet engines and capable of carrying more than 500 passengers.

This dramatic progress in aircraft is due to the rapid development of many technologies in this period of time. New materials that could be used in the frame and covering of the aircraft were developed. Smaller, lightweight engines were developed. The design of airplanes changed to make them more streamlined so they would move through the air with less drag. When combined, these improvements made it possible to build larger, heavier, longer range aircraft. It is important to understand that no single advance made this possible. It required the combination of many technological advances.

The Major Parts of an Airplane

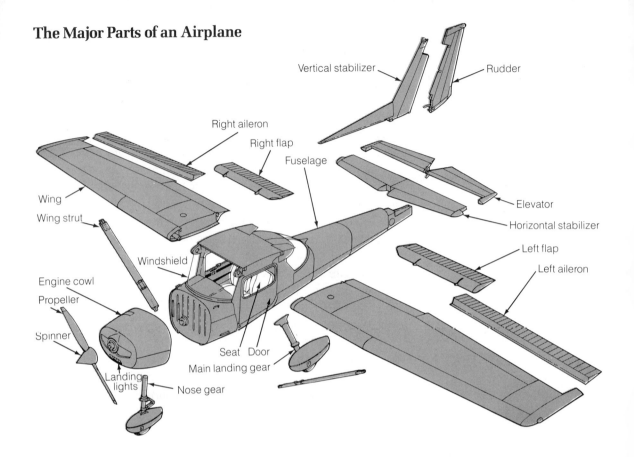

Parts of an Airplane. First let us look at the major parts of an airplane, then break these into smaller parts for study.

The fuselage is the body of the airplane. It contains the passenger cabin, the cargo space, a cabin for the pilot and copilot, and in larger aircraft, space for the navigator, flight engineer, and other crew members and passengers.

The fuselage is very carefully designed. On most single engine aircraft, the engine is in the fuselage. Space provided for the crew and passengers must be comfortable. Passengers in large commercial airplanes must have room to move. High-altitude aircraft must have a sealed cabin for crew and passengers. The cabin must be pressurized and be supplied with air and heat.

The fuselage also holds most of the navigation, radio, and other equipment. In passenger airplanes, equipment must be provided for preparing meals for passengers. Restrooms must be provided. Airplanes that carry passengers long distances often show movies to the passengers.

The *wing* of an airplane provides the lift necessary to keep the airplane airborne. Control surfaces on the wing help guide the flight of the airplane. They operate mainly in turning the airplane left and right. Brakes and flaps on the wing help control the speed of the airplane. In some airplanes, the wing also holds fuel tanks and space for the landing gear to be stored while in flight.

The *tail assembly* is the rear or tail of the airplane and is mounted on the rear portion of the fuselage. It is needed to provide stability to the airplane. The major control surfaces of the airplane are on the tail assembly.

The rudder is on the vertical fin. This is the main left-right control surface. The elevator is located on the horizontal stabilizer. When moved, the airplane moves up or down. Some airplanes are equipped with a stabilator—the stabilzer and elevator made into one unit. It is turned up or down to control the airplane.

The basic *controls* of an airplane are used to change the speed and direction of the airplane in flight. Speed of the airplane is controlled by changing the speed of the engine. The throttle is used to control the engine's speed.

The direction of the airplane in flight is changed by the control surfaces on the wings, the rudder and the elevators. This will be explained in more detail when the various parts of the airplane are discussed in more detail.

How Are Airplanes Built? In nearly all airplanes built before 1930, the airframe was made of wood or steel tubing. A truss design similar to that used for bridges was used. The frame was braced with tie rods and cross members to give strength. Cotton fabric was stretched over the frame and painted or *doped*. This gave a smooth, weather-tight outer skin for the airplane. All parts of the airplane—fuselage, wings, and tail assembly—were made this way.

Some small airplanes are still made in this way. These are used primarily for sport flying and for the first flight instruction a pilot receives.

Most modern aircraft are built using aerodynamically shaped sheet metal panels. The panels are welded or riveted together to form the main structure of the airplane. Only a simple skeleton of framework is used under the panels. The panels are the outer skin of the airplane and give the strength needed. Little internal bracing is needed because the outer skin is strong.

Modern metals are used in these airplanes. Weight is important in airplanes. So, modern lightweight, but strong metals are used.

This Piper Cub was constructed using a tubular truss frame.

Transportation

Types of Fuselage and Wing Construction

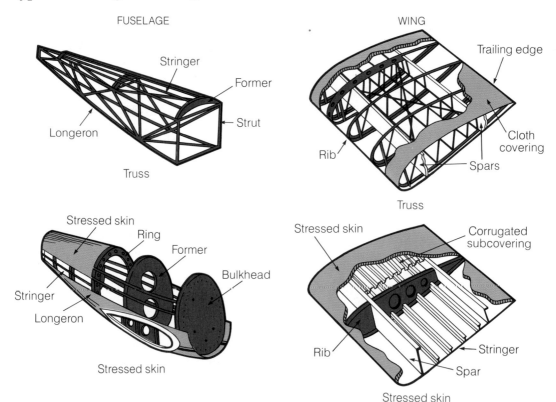

Aluminum and its alloys are the most used metals in aircraft construction. Some magnesium, titanium, and stainless steel are used where there are special needs for these metals. New composite materials are being used to give strength and light weight.

The wing is constructed of spars and ribs. The spar is a long beam-like piece that runs the length of the wing. The ribs are attached to the spar. The ribs give the wing its shape. A covering is then placed over the ribs to form the outer skin of the wing. Multiengine aircraft have engines mounted on the wings. Landing gear often is attached to the wings also.

The tail assembly is constructed in a similar manner to the wing. The tail section does not provide lift for the airplane, so it does not have to be shaped like the wing. The rudder and aileron are usually just flat with the outer skin covering.

How Does an Airplane Fly? Before an airplane can fly, several forces of nature must be overcome. First, the force of gravity must be overcome.

Vertical structures called winglets added to the ends of the wings improve aerodynamic efficiency.

Forces on an Airplane and on an Airplane Wing

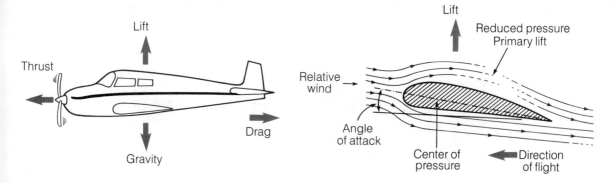

The wing provides lift that overcomes gravity. As the airplane moves through the air, the air flows over the wing. The wing is flat on the bottom and curved on top. When the air moves over the wing, it has to move faster over the top because there is a longer distance to move. This causes a slight low pressure on top of the wing. The wing is also tilted slightly up on the front. This is called the angle of attack. The air hits the bottom of the wing, also causing some lift. The combination of the low pressure on the top of the wing and the air hitting the bottom of the wing produces the lift needed to overcome gravity. The lift must be greater than the force of gravity on the airplane before it can become airborne. Once lift is greater than gravity, the airplane can rise off the runway.

Drag also must be overcome before an airplane can fly. Drag is the resistance created by the air as the airplane moves through it. Airplane designers have made airplanes streamlined to overcome drag as much as possible, but it can never be eliminated completely. The engine produces thrust, or the forward moving force to overcome drag. In a propeller-driven airplane, the propeller pulls into the air causing the airplane to move forward. In jet-powered airplanes, the thrust of the jet engine overcomes drag.

Helicopters. Helicopters are commonly used rotary-wing aircraft. These craft can fly in any direction: forward, backward, left, right and can even hover in one position. This unique movement makes helicopters valuable for many uses.

Small helicopters are used for traffic control, to transport crash victims to hospitals and even in fighting fires. Large helicopters have been built that can lift several tons of cargo.

Helicopter Lift and Thrust

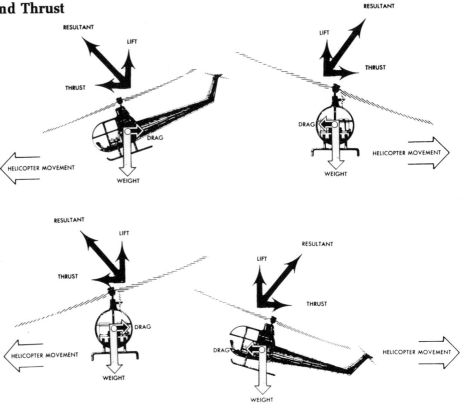

The whirling rotors provide the lift needed to offset gravity. As the rotors move, air flows over the airfoil-shaped blades. This creates some lift. Tilting up the leading edges of the blades to increase the angle of attack increases air pressure on the bottoms of the blades. More lift is created. The pilot can control the amount of lift by controlling the amount the rotor blades are tilted.

A small rotor in the tail of the helicopter is needed to offset the torque of the engine. Without the tail rotor, the fuselage would spin as the rotors spin. By controlling the tail rotor, the pilot can control the direction of flight.

Space Vehicles. There are four categories of space flights. These require different types of *launch vehicles* to carry the *payload,* people and instruments, into space. The four types are: (a) sounding missions, which investigate the upper atmosphere from earth to 6,400 kilometers (4,000 miles) into space; (b) space probes, which go beyond 6,400 kilometers (4,000 miles) into space; (3) orbital missions, which circle the earth, with the

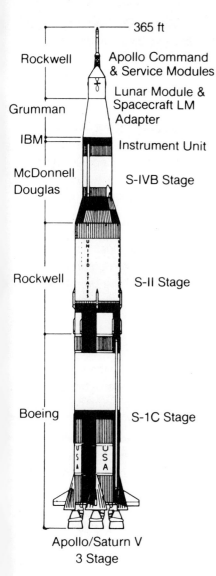

Apollo/Saturn V
3 Stage

spacecraft reaching speeds of 28,000 kph (17,500 mph) to 39,360 kph (24,600 mph); and (d) deep space probes, which reach for other celestial bodies in space such as our moon and Mars. The payload a space vehicle carries and what the mission will be determines the launch vehicle to be used.

The Scout is the smallest and simplest launch vehicle used in the United States. It uses solid-fuel rockets on all four stages. Solid-fuel rockets are simple to operate and can be stored for a long time. Once ignited, they cannot be turned off and restarted. The Scout can carry a payload weighing 57.6 kilograms (240 pounds). It is used for unmanned missions for probes into space, for launching small satellites, or for sounding missions.

Larger payloads require larger and more powerful launch vehicles. The Atlas-Agena can lift 2,778 kilograms (5,950 pounds) into earth orbit, or lift 427 kilograms (950 pounds) to the moon. This launch vehicle is a two-stage, liquid-fuel rocket.

The largest U.S. launch vehicle is the Saturn V, which can lift a payload of 633,000 kilograms (285,000 pounds) into orbit. Nearly 44,100 kilograms (98,000 pounds) can be carried to the moon. The Saturn V was used to launch the American moon flights from earth. It was designed for the Apollo moon missions and for other interplanetary exploration. As with most of the modern launch vehicles, the Saturn V is a multistage rocket. Separate stages are used so that empty fuel sections can be dropped after the fuel has been burned. With the first stage of a rocket dropped, the remaining rocket weighs less and requires a smaller engine and less fuel. The first stage must be the largest and most powerful because it must lift the entire rocket from earth and start it on its way to space. Gravitational forces must be overcome. Once into outer space, only a small amount of thrust is needed to move a spacecraft. There is no gravitational force to overcome and the craft is weightless.

Power Plants

In this discussion of power plants used in transportation vehicles, little mention will be made of how the engines operate. This has been covered in Chapter 3, Energy and Power. Refer to that chapter as you read here about the various power plants.

What is important is that you understand the various power plants and why they are used. The power plants used in transportation vehicles are special. The engine used in an airplane will not work in a ship. A rocket engine is not practical in an automobile. Each power plant is designed for a specific use.

Diesel engines are used in automobiles, ships, and trains. But, these engines are specially designed. The diesel in the automobile is designed specially for that use. The diesels used in the ship or train are designed specially for those uses. Can you imagine trying to put a diesel train engine in an automobile?

Automobiles. The automobile engine is the best known of all the power plants used in transportation vehicles. Most automobiles in use today have an internal combustion engine for power. Most of these are gasoline fueled. Some are equipped with a diesel engine. Chapter 3, Energy and Power, explains both these internal combustion engines.

The first automobiles were powered by steam engines. These were slow, noisy, dirty, and inefficient. The earliest automobiles had to stop every ten or fifteen minutes to build up steam. They discharged a lot of smoke and ashes from the fire box. Possibly the most famous of the early steam cars was the Stanley Steamer. It was built from 1897 to 1924.

Early automobiles were also powered by electricity. During the late 1890s and early 1900s the electric car was very popular. These were quiet, clean, and smooth running. But they could only travel a few miles before needing to recharge the batteries.

Both the steam engine and electricity still are being considered as power sources for automobiles. Many people feel the electric car is an answer to the oil shortage, but the problems with early electric cars still exist. Even the modern electric cars need recharging after only a few miles of travel. The batteries take a lot of space in the car. The batteries also are expensive.

This full-scale model of a three-wheeled urban car is 10 feet long. The canopy opens up like a small jet plane.

The internal combustion engine as we know it today was developed in the late 1800s. Since that time these engines have been used in automobiles. Charles E. Duryea and his brother, Frank, are considered to have built the first successful American gasoline-powered automobile. In 1894 their auto made its first successful trial run. In 1895 they started the first American automobile manufacturing company.

Since this early start, the automobile engine has been developed into a very complicated machine. The modern automobile engine can propel the vehicle to more than 100 miles per hour. It can run for hundreds of hours with little or no maintenance. The engines used in large trucks generate hundreds of horsepower.

Trains. The earliest trains were steam powered. Wood was used as the fuel for the early locomotives. Then coal became the most used fuel. Steam engines were used until well after World War II. In the late 1930s the greater efficiency of diesel engines made steam less economical. Today only a few excursion trains use the steam engine. Most of these are run by hobby groups such as steam train societies.

In the United States today, all trains use diesel, electric, or gas turbine power. Some countries still use steam locomotives. The steam engine used only 8 or 9 percent of the energy in the fuel. Diesel locomotives can be as much as 40 percent efficient. This means large savings in energy. Steam engines can operate at maximum power for only a short time. Diesel engines can operate at maximum power indefinitely.

Some diesel locomotives have the engine connected to the wheels through a transmission system. Some are similar to the transmission in an automobile. Other locomotives are called diesel-electric. The diesel engine turns electric generators. The generators produce electricity. This is used to run electric motors on the axles of the locomotive. Thus the diesel engine is used only to produce electricity. The electricity is used to run the motors to turn the wheels of the locomotive.

Some locomotives, used mostly in West Germany, use diesel engines to turn hydraulic pumps. The pumps are connected to

Electricity from the overhead wires powers this train.

hydraulic motors on the axles. These motors provide the power needed to move the train.

Gas turbine engines also are used on trains. These are similar to the turbine, or jet, engines used on aircraft. Though not used widely, the gas turbine engine may be the power plant for trains in the future.

Some trains do not have an engine. Have you ever seen a commuter train that had no engine and wondered what pulled it along the tracks? Many commuter trains use electric motors for power. The motors are placed on each axle of each car. The motors are connected to overhead electric lines. In this way each car contains its own power plant. This makes it possible to add or subtract cars without changing locomotives. Each car can be operated separately or joined to make up a train.

The electric cars can run in either direction. The train does not have to be turned around for the return trip. The electric train needs only a small crew to operate. An engineer usually is needed to start and stop the train and to run it at the proper speed. There is usually a conductor for each car to take tickets and to check passengers.

Freight trains still have the separate locomotive. The freight train is made of many cars. Many of these are special-purpose cars. To put electric motors on each car would be quite expensive. Often the cars sit for long periods of time waiting to be loaded.

Ships. The power plant used in a particular ship depends on many things. One of these is the cost. A shipbuilder must not raise the cost of the ship too high. If the cost is too high, the buyer might not buy it.

The speed of the ship is important. If a ship is being designed to move at high speed, a high-speed power plant must be used. The fuel burned in the engine is also important. Often a ship is designed to travel a certain route. There must be fuel for the ship's power plant along this route. If several different kinds of fuel are available, it would be more economical to use the cheapest kind of fuel. The power plant in the ship would be designed to burn the cheapest fuel.

The size and carrying capacity of the ship is important. The power plant must be strong enough to move the ship when loaded with cargo. It must do this as economically as possible. The power plant also must run with low-maintenance cost. If many repairs are needed, the cost of operating the ship would be high.

In 1914 it was estimated that 90 percent of the merchant ships used coal as fuel. Today very few burn coal. Oil is a much easier and cleaner fuel to use. The old coal burning ships required large gangs of men to feed the boiler. The coal took up large spaces in the hold. This cut down on the amount of cargo that could be carried.

Burning coal was dirty. Coal dust went all through the ship. Ashes had to be removed from the boilers. The ashes then had to be discarded.

Oil is much cleaner. No gangs of men are needed in the engine rooms. In fact, some modern ships can operate for long times without anyone in the engine rooms. An engineer can stay on the bridge of the ship and monitor gauges that tell the condition of the engines.

Fuel oil takes up less space than coal. This leaves more space for the cargo. By carrying more cargo, the shipper can make more profit from each trip.

The power plants in most modern ships are powered either by diesel engines or steam turbines. The steam turbine operates smoothly. It is also reliable and can operate a long time with few repairs. But the steam turbine is large. It takes a lot of space in the hold of the ship. It is used in some large passenger liners because space is not as important. The steam turbine also gives faster speeds. This also makes it popular for use in passenger ships.

The diesel engine is used more and more. It uses less fuel, which means less space is needed for fuel tanks. Over one-half of modern merchant ships are diesel powered. Nearly all ships on the inland waterways are diesel powered.

Some modern ships are diesel-electric, or turbo-electric powered. The power plant (diesel or turbine) is used to turn a generator. The electricity runs electric motors coupled directly to the propeller shaft.

A few ships are nuclear powered. The USS *Savannah* was the world's first nuclear-powered merchant ship. It was launched in 1959. But, nuclear ships are costly to build and to operate. For this reason they have had little use as merchant ships. Some warships are nuclear powered. Nuclear powered submarines can stay under the water for long periods of time. They do not have to come to the surface for frequent fueling.

Airplanes. The power plant is the heart of any airplane. Without some source of power to overcome the forces of gravity and drag, the airplane would not be able to fly. Unpowered flight was possible long before the Wright Brothers flew their airplane. But the engines in these early days were heavy. They did not produce enough power to overcome their own weight.

The design of airplanes is affected by the engine being used. In many cases the airplane is designed around a desired engine. The first airplane engines were inline piston engines. Look at the styling of early airplanes. You will see the effect of the shape of the engine. The inline engine was rather long and slender. The pistons in the engine were lined up one behind the other. This allowed rather streamlined fuselage for the single-engine airplane.

In 1925 the Wasp radial piston engine was developed for airplanes. It provided 400 horsepower. With these engines in airplanes, the fuselage was made more round in shape. This was the shape of the engine. The fuselage was made larger and more round in front to enclose the engine.

Jet engines are long and slender, much like a cigar. This allows a long and slender fuselage. Look at the first jet-powered airplane. Compare their design with the propeller driven airplanes at the same time. You will see that the jet engine allowed a rather dramatic change in airplane design.

As more powerful engines were built, larger airplanes that could carry larger loads could be built. By using several engines, even heavier loads could be carried in the airplanes designed to carry cargo.

The different kinds of aircraft engines will not be discussed in detail here. Refer to Chapter 3, Energy and Power, for a more detailed discussion of airplane engines.

It is important to understand that the engines used in airplanes are special engines. Airplane engines must meet strict specifications and must be able to operate in ways that other engines such as automobile engines don't. An airplane engine must be reliable. It must run well for long periods of time without failure. It must operate in different positions, even upside down or on its side. The engine must operate in extreme weather and temperatures. Unlike an automobile, it is not possible to pull to the side of the road and wait for a tow truck when an airplane engine stops running.

Piston engines and jet engines are used in modern airplanes. Most large commercial airplanes are powered by jet engines. The jet-powered airplanes can fly faster and can carry larger loads for longer distances. This is important in commercial transportation.

A *Double Wasp* radial airplane engine with two rows of cylinders, one behind the other.

But have you wondered why different engines are used in different airplanes. Why not have just one type of engine that is used in all airplanes? The reason is quite simple. Different kinds of airplanes are needed, just as different ships are needed. Airplanes are designed to do different jobs.

Some airplanes are designed for small airfields. Some are designed for sport flying. The military needs airplanes that are extremely fast. The commercial airlines need large passenger and cargo aircraft. One engine cannot do everything well.

At speeds below 560 kph (350 mph) a propeller engine is very efficient. As speed increases above that, the efficiency drops off. Jet engines are just the opposite. The rocket-like action of a jet is more efficient at higher speeds.

The turbo-jet is very inefficient at low speeds. At 480 kph (300 mph) only half the power produced by the engine is used for thrust. The rest is used to turn the turbine in the engine.

One can see that neither the propeller nor the pure jet engines operate in the medium speed ranges of 560-720 kph (350-450 mph).

One solution to this has been the turboprop engine. (See Chapter 3, Energy and Power.) This engine provides the power to turn a propeller. The propeller provides the thrust. The turboprop engine is efficient in medium speed range.

The turbofan engine has become popular because it is efficient at speeds above 320 kph (200 mph). This engine combines the thrust of the jet and the propeller. There is less wasted energy. Most of the power produced goes to produce thrust. Some of the turbofan engines are very large. One has a fan over 3 meters (10 feet) in diameter.

Now you should be able to see that different engines are used for different needs in airplanes. The selection of the engine for a particular airplane is important. One would not put a gasoline engine in a model airplane designed to be powered by a rubber band. Likewise, one would not put a turbojet in a slow speed airplane.

Spacecraft. Just as with airplanes, the power plant of a spacecraft is the heart of the craft. Without it, the flight would not be possible. The power plant must provide enough power to overcome the forces of gravity and drag before the craft leaves the earth. In spacecraft this force, the thrust of the engines, must be great enough to lift the vehicle from the earth and move it out of the earth's gravity. The thrust of the engine must be greater than

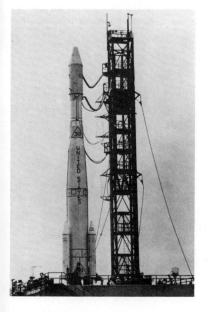

A multistage rocket ready for launch.

Upper Stages of Rocket Booster

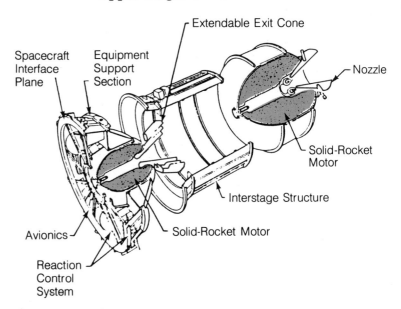

the total weight of the vehicle. See Chapter 3, Energy and Power, for a more detailed discussion of rocket engines.

Most modern rockets have several engines that fire in stages. Stage one is the largest and most powerful because it must lift the entire rocket from the earth and begin pushing the rocket into space. The first stage uses all its fuel and drops away. The second stage then fires and continues to push the vehicle into space.

The second-stage engines are smaller. They do not have to push the weight of the first stage so less weight must be overcome. As the rocket moves farther away from the earth, the force of gravity is less also.

Some rockets such as the Saturn V launch vehicles have a third stage. It is smaller than the first and second stages and is used to push the vehicle farther into space.

The Space Shuttle represents a new type of launch vehicle. It is reusable. It consists of a reusable orbiter mounted on a large fuel tank and two solid-fueled booster rockets which can be recovered and reused.

When launched, the two solid-fuel rockets and the three engines on the Space Shuttle orbiter burn at the same time. The solid fuel rockets burn out and are parachuted into the ocean. They are picked up by a recovery ship for reuse on other launches. The large fuel tank continues into space with the orbiter. When it is

Family of Current U.S. Launch Vehicles

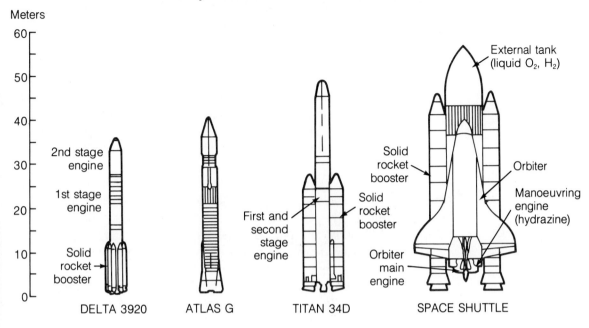

empty, it is dropped away and falls back to earth, landing in the ocean.

The orbiter vehicle then remains in space to finish its planned mission. When the mission is completed, the orbiter returns to earth, landing like an airplane. Each orbiter is designed so that it can be used for many missions. It takes several weeks for the returned obiter to be prepared for the next flight.

The big advantage of the Space Shuttle is that it is reusable. In the past all space vehicles were designed to be used only one

The Space Shuttle Orbiter *Enterprise* after landing.

time. This is the same as having to buy a new car each time one drives to the grocery store! The cost of each trip would be extremely high if one had to buy a new vehicle for each trip. With a reusable vehicle the cost of the vehicle is spread over many trips, making the cost per trip much less.

The Space Shuttle replaces many of the expendable vehicles once used. It carries civilian and military payloads, including those of some other countries. Shuttle crews are able to service and repair satellites in orbit. Satellites that are not working can be retrieved and then returned to earth. It is possible to perform rescue missions in space.

Space transportation is still mostly experimental. Regularly scheduled cargo or passenger flights into space are still far in the future. They still are limited to fantasy in books or on television.

Though few people have travelled in space, millions have benefited from the space program. There have been many spinoffs from space research.

Weather information from satellites is among the best known benefits from the space program. Pictures of cloud formations over the United States are used by weather forecasters. The weather reporter on a television news program uses the weather satellite pictures. Other benefits of space research are discussed in Chapter 9, Consequences of Technology.

Controls
To help better understand controls on transportation vehicles, let's first look at the bicycle. This is a transportation vehicle you probably know well. What controls are on the bicycle? How does one control a bicycle?

First, one can control the direction it travels. This is steering. You steer the bicycle by moving the handlebars. They are connected to the front wheel. Move the handlebars to the left. The front wheel turns left. This makes the bicycle and rider turn to the left. Move the handlebars to the right and the rider turns to the right. Most transportation vehicles have some way of steering or controlling their direction of travel.

Speed of travel must be controlled. Too much speed can cause an accident. Travel with too little speed takes too long to make a trip. On the bicycle, one controls the speed by pedaling. Pedal quickly and the bicycle goes faster. Pedal slowly and the rider goes slowly.

The brakes on a vehicle are part of the speed control. They slow the vehicle or stop it. On the bicycle the brakes may be the coaster type that are made into the rear axle. Many modern bicycles have hand brakes. These squeeze pads against the wheels to slow the speed.

Automobiles. Automobile controls are similar to those of the bicycle. Both direction and speed of travel can be controlled. But the controls on an automobile are a bit more complicated. To steer the automobile, the driver moves the steering wheel. This is connected to the front wheels. Turn the steering wheel left, and the car turns left. Turn to the right, and the auto moves right.

Many modern automobiles are equipped with power-steering systems. A hydraulic pump is connected in the system. When the steering wheel is turned, the pump is activated. Hydraulic fluid is pumped to cylinders on the front wheels. These cylinders push against the wheels to turn them. This makes steering much easier. The driver does not have to exert as much force on the steering wheel.

Like the bicycle, the speed of an automobile must be controlled. The driver controls speed by controlling the speed of the engine. Pushing on the throttle (gas pedal) causes the engine to run faster. The engine usually is connected to the rear wheels. So, when the engine runs faster, the rear wheels turn faster. The

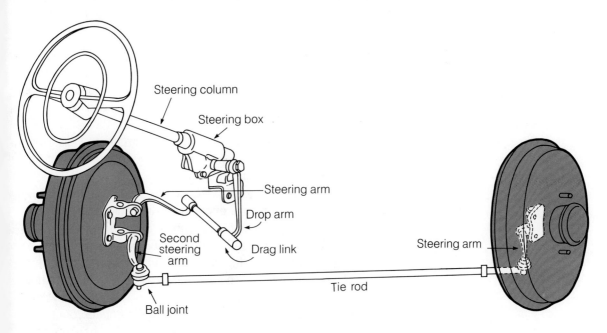

Steering System of an Automobile

engine is connected to the wheels through the drive train. The engine and drive train are the power system of the automobile.

Like the bicycle, the automobile also must have brakes. The braking system of the automobile is made of the brake pedal, the hydraulic system of pumps and lines, and the brakes. To apply the brakes, the driver presses on the brake pedal. This causes the hydraulic pump to press liquid to the wheel cylinders. The pressure in the cylinders causes the piston to exert pressure on the brake. The brake shoes or pads are pressed against the wheel. The friction between the brake shoes or pads and the wheels slows the automobile. The braking system of an automobile is illustrated in Chapter 3.

Trains. The control of trains is rather simple. The direction of the train is controlled by the track. The train wheels are rimmed or flanged. The flanges keep the train wheel on the track. The engineer does not have to steer the train. It can travel only where the tracks are.

The engineer can control the speed of the train. Like the automobile, the train has a throttle, which controls the speed of the engine. If the throttle is opened, the engine runs faster. This makes the train move faster. Close the throttle and the train slows. The engineer also can apply brakes to slow or stop the train.

Ships. Control of the direction of travel of ships is similar to airplanes. The ship travels in water, the airplane in air. Both use a rudder to control direction of travel. In the ship, the captain turns the wheel to move the rudder. The rudder is located in the stern of the ship below the water line.

When the rudder is turned left, the ship turns to the left; when turned to the right, the ship moves right. To travel in a straight line, the rudder is held straight.

Like other vehicles, the speed of the ship is controlled by changing the speed of the engines. In the ship, the engine is connected to the propeller. The propeller turns in the water. This exerts a force to push the ship through the water.

In many smaller ships, the captain moves a throttle to change the speed of the engines. This is similar to pushing the throttle of the automobile. Some large ships use a telegraph system. Through this system the captain signals to the engine room what speed he wants. The workers in the engine room then speed or slow the engines. In early steam-powered ships, the people in the engine room had to increase steam pressure to increase engine speed. This often meant adding more coal to the boilers to make the fires hotter.

Many modern ships are automated. The speed of the engines is controlled by computers. The captain moves a throttle, and the engine changes speed. No people are needed in the engine room to operate the engines.

Ships do not have braking systems like the automobile. Some braking action can be had by reversing the engines. This reverses the propeller. The propeller moving in reverse pulls against the movement of the ship. Even with the propellers reversed, today's huge ships may travel several kilometers (miles) before they are stopped completely. The captain must plan far ahead.

Airplanes. Earlier in this chapter you learned how an airplane flies, but that is only a small part of the flight. The pilot must have some way to control the direction of flight. This is done by moving the control surfaces on the wings and tail assembly. The ailerons on the wings control the roll of the airplane. *Roll* is an airplane's movement as one or the other wing tip dips. As one wing tip dips down, the other raises. The pilot controls roll by turning the wheel or, in a pilot's term, the *yoke*. On some smaller airplanes the pilot controls the roll with a *stick*. To roll left, the pilot turns the wheel or yoke to the left. This causes the aileron on the left wing to raise and the aileron on the right wing to lower. The left wing is forced down and the right wing up. The airplane rolls left. On a stick controlled airplane, the stick is moved to the left to make the airplane roll left. To roll right, the wheel or stick is moved to the right.

Yaw is the movement of the airplane right and left. The pilot controls yaw by moving the *rudder pedals*. This moves the rudder left or right. If the rudder is moved to the right, the force of the air on the rudder forces the airplane to turn to the right. A left rudder causes the airplane to turn to the left. Thus, if the pilot wants to turn the airplane to the left, he presses on the left rudder pedal and the airplane turns in that direction. A right turn is made by pushing on the right rudder pedal.

Actually, a turn in an airplane is made by altering the surface of the wing—the aileron and the surface of the vertical stabilizer, and the rudder, moving both the yoke and the rudder pedals. If only the rudder were used to turn the airplane, the plane would skid around the turn thus making a flat turn. To prevent this, the pilot rolls the airplane slightly, using the yoke to move ailerons while

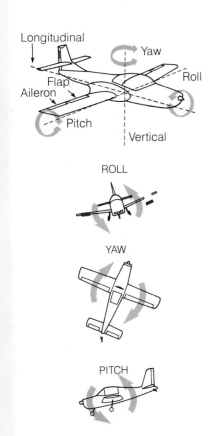

Movements of an Airplane in Flight

moving the rudder with the rudder pedals. So, to make a turn to the left, the pilot rolls the airplane slightly to the left and moves the rudder left, at the same time. With practice the pilot can make a smooth turn.

The third control of the airplane is *pitch*, the movement of the airplane upward or downward. Pitch is controlled by the elevators. When the elevator is raised, the force of the air causes the tail of the airplane to be forced down and the airplane is pitched upward. If power is added, the airplane will climb. When the elevator is moved to a down position, the tail is forced up and the airplane dives.

Controlling Flight

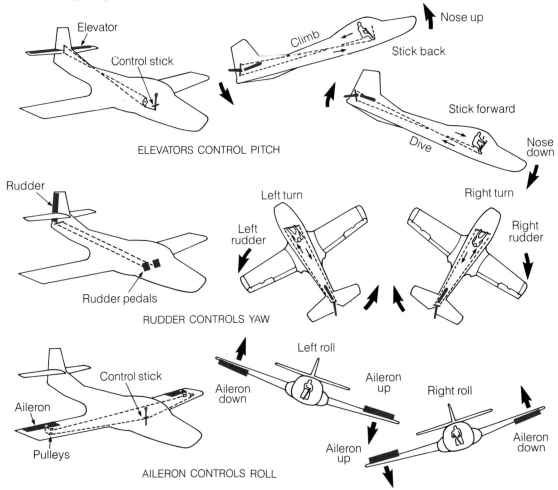

The pilot moves the elevators to control pitch by pulling or pushing on the yoke or stick. When the stick or yoke is pushed forward, the elevator is moved downward and the airplane dives. When the yoke is pulled back, the elevator is moved upward and the airplane climbs.

In actual flight, the pilot is controlling all three of the motions of the airplane to keep it in level flight, or to make turns, dives, or climbing turns or dives. This requires excellent coordination and much practice. Once flying skills are mastered, the pilot must keep in practice by flying frequently.

The Way

The *way* is a term used to describe the surface or route a vehicle travels. The automobile travels on a highway. The train runs on a railway. The ship travels on a waterway (lake, river) or shipping lane in an ocean. The airplane travels in air corridors.

For a more detailed discussion of the construction of highways and railways see Chapter 5, Construction. Our discussion here will not include construction.

Highways. Roads and highways make up a system of routes for automobile travel. The modern super highway has grown from the trails of foot and horseback. Many modern roads follow the trails of early foot travel.

The modern automobile requires a smooth highway. In the early days of automobiles, the roads were only cleared strips through the forests and fields. In dry weather they were dusty. When it rained, they were muddy. The corduroy road, which was surfaced with logs, was rough and bumpy. As the automobile developed, more and better roads were needed.

Primary highways are the major highways. They carry traffic between larger cities. The Interstate Highway System is part of the primary highway system. These are controlled access highways. Motorists can enter or leave only at certain places. These places are usually located at intersections, where other highways cross.

A cloverleaf is a large intersection of two or more highways. It allows traffic to enter or exit without stopping or slowing other traffic. Some cloverleaf intersections are built with several levels of traffic crossing each other.

The complete highway system is more than just roadway. Bridges are needed to span rivers and lakes. Tunnels carry traffic through mountains. Overpasses allow highways to cross without an intersection. Rest areas are needed so motorists can stop regularly on long trips.

A modern highway also has several controls to regulate traffic. Speed-limit signs tell the driver the safe speed. Information signs tell of services such as food, fuel, or lodging ahead. Other signs tell of upcoming exits.

All these, the highway, bridges, tunnels, and signs are part of the highway system. Without this system, modern highway transportation would not be possible. If one wants, one can drive more than 4,800 kilometers (3,000 miles) across the United States virtually nonstop. A drive across the city is possible on the city streets. A drive through the country is possible on the secondary roads and country roads.

Railways. The railway (tracks) makes rail transportation different from all others. The track controls the vehicle's direction of travel. In all other vehicles, the driver of the vehicle has this control.

Most railways are made by first laying a roadbed of stone and gravel. Wooden sleepers, or ties, are laid on the bed. The rails are fastened to the sleepers. Some more modern railways have been made using concrete sleepers. Others have been made with a concrete roadbed. The use of concrete gives a smoother ride for the train.

A rough or uneven railway can cause the train to derail. As the train moves over uneven sections of track, it moves from side to side. If this motion is too violent, the train will derail.

The railway is a system. It is made up of the way including the roadbed, various kinds of yards and terminals, and a system of signals and controls. It also involves switches to move the train left or right onto other tracks. Bridges are constructed over rivers, streams, and deep gulleys. Tunnels are dug through mountains. The train cannot go up a steep grade like an automobile. The grade must be gradual. To get to the other side of a mountain, train tracks must either be laid around the mountain or through it in a tunnel. Overpasses are built to allow the train to cross highways without stopping traffic. Crossings are also built where highways are intersected without an overpass. Here signals are installed to warn motorists.

Classification yards are large areas where freight cars are sorted. The cars are moved into tracks to place them in a specific order. The train cars are then coupled together to form the train. Train cars are repaired and cleaned in service yards.

right In a classification yard, train cars are electronically sorted and placed in the proper order to make up a train. *below* The controller in a tower overlooking the classification yard.

The terminal is a place for passengers to wait for trains. Some large terminals also have restaurants and gift shops for the waiting passengers. At one time train terminals were large and very elaborate. Today most have closed. Few trains carry passengers. Passenger service is only a small part of train traffic.

Many people feel the monorail is the railway of the future. Monorails have several advantages. Most are built above ground. They are built on posts or towers. The train vehicle is suspended under the track. These systems can be built in cities without demolishing many buildings. These elevated railways are much cheaper to build than subways, though both are expensive.

A system of signals is used to help control trains. These are usually lights that tell the engineer conditions ahead. They may tell what speed to travel. They tell if there is another train on the tracks ahead. If so, the engineer may have to move the train to a side track. After the oncoming train passes, the engineer moves his train back onto the main track.

Airways and Waterways. The way for air and water transportation is quite different from land. On land, the vehicles are limited to a constructed way. On water or in the air there are no limits. This does not mean that airplanes or ships travel at random. They travel in set lanes. These are called *sea lanes* and *air corridors*. Many are set by international law or agreement.

For example, there is heavy traffic in the English Channel. Routes are set to reduce the risk of accidents. In one 24-hour period as many as 800 ships may pass one point in the channel.

Working in Transportation

Civil Aviation

In 1983 about 398,000 people in the United States worked in air transportation. Almost 70 percent of these worked for commercial airlines. Of these persons only about 20 percent worked as pilots and flight attendants. The rest worked in ground jobs. These included mechanics, cargo handlers, air traffic controllers, and reservations and ticket agents.

Jobs are available in civil aviation to persons with a wide variety of training and experience. Pilots must be highly trained and are usually required to have a commercial or air transport pilot's license before they begin working for an airline. They usually begin as a flight engineer and work their way up to a co-pilot and then pilot. Physical requirements for pilots are high. College graduates are preferred.

Other jobs in aviation do not require such high levels of education or physical standards. Flight attendants with some college training are preferred. Mechanics should have high school or trade school diplomas. Experience as a mechanic is helpful.

A high school education is also required for reservations, ticket and passenger agents. These persons must have a pleasant personality and good speaking voice.

Working conditions in civil aviation are generally considered to be desirable. Airlines operate flights day and night. Some workers have to work irregular hours. Flight personnel travel long distances and are often away from home as much as one-third of the time. Salaries are above average for persons working in private industry.

Merchant Marine

In 1983 about 15,300 persons in the United States worked on merchant vessels. 229,000 persons worked in American shipyards in

1983. Most of these people repaired ships. Many of these jobs are highly specialized and are found only in this industry. Over 60 percent of the U.S. merchant marine fleet are cargo vessels. There is only a small fleet of passenger ships.

More than one-half of the workers in the merchant marine industry make up the ship crews. Most of the workers on shore are dock workers—loading and unloading cargo, and maintenance, for example. A small percentage are clerical and administrative workers.

Work in the merchant marine is varied, as ship crews require people with special skills. For many, a high school diploma is not required but as with most any job, a high school diploma is preferred. Advanced education is an advantage and usually necessary for positions such as engineers, navigators, and ships officers.

Persons who work at sea usually must be in good physical condition. They must be able to work as part of a team. They must also be able to work under some military-like discipline.

Working conditions can be strenuous and even hazardous. Ship crews may have to work long hours, usually working split shifts around the clock. Dock workers usually work a 5-day, 40-hour week.

Earnings on board ship are relatively high. However, annual income for both shore-based and ship-board personnel can be affected by irregular shipping schedules.

Railroads

Railroad workers can be grouped into four main groups: operating employees (train crews), station and office workers, equipment maintenance workers, and property maintenance workers.

Train crews make up about one-third of all railroad workers. They include engineers, conductors, and brake operators. They work as a team to operate the trains.

About one-fourth of all railroad workers are station and office workers. These include management personnel, agents, baggage and freight handlers and persons in the various business jobs.

The remaining rail workers are maintenance workers. They repair the trains and the tracks, tunnels, signals and buildings.

Most railroad workers are trained on the job. Workers, such as those on train crews, must be in excellent health. Promotion is usually based on seniority and ability. Earnings of railroad employees vary widely depending on the work done: Most work a 5-day, 40-hour week.

Trucking

About three-fourths of all trucking workers have blue-collar jobs, including truck drivers. Others are mechanics, freight handlers, and supervisors. The remaining one-fourth are white-collar workers and include clerical and administrative personnel.

Most of the blue-collar workers are hired as unskilled workers. The work is learned on the job. No formal education is required. Many employers prefer to hire people with high school diplomas. In most jobs in the trucking industry good physical condition is necessary.

The administrative jobs are usually filled by college graduates. Clerical workers have usually completed some clerical training in high school or trade school.

Earnings are relatively high in the trucking industry because the higher paid drivers make up a large proportion of the workers. Working conditions vary widely depending on the kind of work done.

Opportunities in Transportation

The number of workers needed in transportation is expected to grow about as fast as the average for all industries. More workers may be needed in civil aviation. A decline is expected in merchant marine workers.

Energy concerns could have an impact upon all mass transportation. If people use their personal automobiles less and mass transit, such as trains, more, the need for workers in mass transit could increase.

Suggestions for Further Study

Personal Involvement

There are many hobby areas that can help you learn more about transportation technology. Many of these are the type of activities you may want to do at home. Many hobby clubs work in these areas. Is there one in your community?

Model airplane building
Model ship building
Model railroading
Model rocketry
Bicycle maintenance and repair

Questions

1. In what ways do modern people depend upon transportation?
2. How has progress in transportation helped (or held back) other areas of technological progress?

3. What are the components that make up a transportation system? Why is each important?
4. How is modern ship building different from 100 years ago? 200 years ago?

Key Words

vehicle	conveyor
power plant	navigation
controls	general aviation
the way	pipeline
tramp freighter	escalator
LASH ship	gondola
tanker	caboose
barge	fuselage
dry dock	rudder
shipway	aileron
transportation	horizontal stabilizer

Designing an Egg Passenger Vehicle

Concept: *Vehicle Safety*

In transportation the most precious cargo is the human passenger. Each year thousands of people are killed and injured in automobile accidents. What can be done to reduce this injury and death?

Industrial products are designed so they serve a particular need. Sometimes this need is for looks or style. Often it is a combination of style and function. The product is designed to do a particular job. In transportation vehicles that job is to carry a cargo.

Objective

In this activity your problem is to design a safety vehicle. Your precious cargo will be an egg passenger. The four-wheeled vehicle will roll eight feet down an incline into a brick barrier. Can you design your vehicle so the passenger will not be injured? Different sized test vehicles will add interest to your test. Which protect better—larger or smaller vehicles?

Procedure

Construction   Construct your test vehicle using materials available in your school or home. No kits are allowed. Remember that in automobiles, the driver and passenger must be able to see out, so, your egg passenger cannot be completely covered. Leave the passenger's "face" showing. Protection for the passenger can be seat belts, padded dash, padded steering wheel, etc.

Can you make your vehicle to scale? Try to make the vehicle close to the scale of an actual automobile. How large is an automobile? How large is the passenger space? What scale space is needed for your egg "passenger"?

Testing   For your vehicle to be judged a safe car, the egg passenger cannot be damaged in any way. The shell can not be cracked or broken. The yolk cannot be broken.

Test Conditions

1. The raceway is constructed from a piece of plywood with one end raised four feet off the floor.
2. A row of bricks is placed at the bottom of the incline. This will be the crash barrier.
3. All cars are placed on the incline with the back wheels over the edge of the plywood. To run the test, the back weeks of the vehicle are lifted onto the incline and the vehicle is allowed to roll down into the barrier.
4. A stop watch should be used to time the roll of the vehicle. Begin timing when the vehicle starts moving. Stop timing when the vehicle hits the barrier. The vehicle should travel down the incline in less than five seconds. If travel time is more than five seconds the vehicle will have to be re-designed.
5. Calculate the speed of the vehicle. If the vehicle travels a distance of eight feet in four seconds, this is how many miles per hour? (1.36)
6. After the test run, the egg is inspected for damage.

The test can be varied by changing the incline of the test ramp. Raising the end of the ramp will increase the speed of the vehicle. This will increase the severity of the impact with the barrier.

How fast can your vehicle travel without damaging the passenger? How can you re-design your vehicle to give more protection to the passenger?

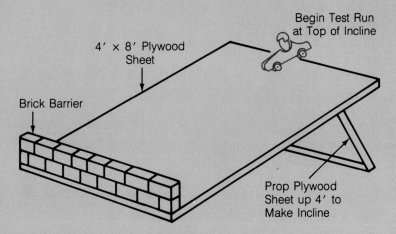

Safety Vehicle Test Incline

Chapter 7   Communication

Introduction

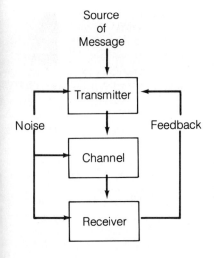

Satellites provide electronic communications between distant points on earth. This one is shown relaying signals from Alaska, Hawaii and the continental United States.

Before we look at communication systems, we need to answer the question, "What is communication?" How does communication take place?

Communication begins with some source of information. When one is talking with a friend, the brain is the source of information. Vocal cords and mouth become the transmitter. Speech or voice is carried through the air, which is the channel. What is the receiver? The friend's ear is. It acts as a receiver and sends the message on to the friend's brain. The brain is the destination. It decodes the information, making it understandable to the friend.

In radio, the announcer is the source of the message. The announcer speaks into a microphone. The message goes into the electronic equipment. This equipment changes the message into a radio signal. The signal is transmitted from the antenna through the air. The radio waves are the carrier of the message. These waves are received by one's radio. The message is then sent through the speaker to one's ears, then to the brain where the message is decoded.

Interference

A sports fan listening to his favorite team on the radio may experience interference. The radio signal may fade away. This interference is a hindrance to the communication process. Other terms for interference are snow, noise, and static. These interferences to communication are found in many media such as television, radio, or recordings.

HISTORICAL DEVELOPMENT OF COMMUNICATION

Time	Development
1456	Johannes Gutenberg prints Latin Bible—first use of moveable metal type
1474	First book printed in English, *Histories of Troye*
1492	Columbus discovers New World
1622	First English newspaper
1639	First press set up in the colonies—Cambridge, MA
1652	First food advertising—a handbill for coffee
1784	*The American Daily Advertiser,* Philadelphia, becomes the first daily newspaper in America
1858	First transatlantic telegraph cable laid—failed after two weeks
1859	The web press comes into use
1865	Sulphite wood pulp process of making paper perfected
1866	First successful transatlantic telegraph cable completed Newfoundland to Ireland
1875	World's first radio operated by Elihu Thomson
1876	Alexander Graham Bell demonstrates telephone
1883	Founding of first strictly women's magazine
1886	Development of halftone process by Frederick E. Ives—begins replacing wood cuts as illustrations
1891	Thomas Edison patents motion picture camera
1892	Color photography patented
1902	Marconi sends wireless message across the Atlantic First motion picture theatre opens in Los Angeles
1906	Lee De Forest invents the audion tube
1913	Parcel post service begins in U.S.
1914	Panama Canal opens
1918	First airmail route in U.S.—New York—Washington—Philadelphia Electron is discovered
1920	First presidential election broadcast on radio
1923	First transatlantic short wave radio
1928	Televison demonstrated
1938	First xerographic copy produced
1939	First sponsored TV broadcast—baseball
1946	**First use of electronic digital computer.** Printed circuits developed
1948	Transistor developed
1953	First compatible color telecast of commercial program
1955	Optical fibers first made—London, England
1956	First transatlantic *telephone* cable completed
1957	USSR launched Sputnik I satellite—space communication was born
1960	LASER developed
1962	Telstar I—First American communication satellite
1965	Sponsored TV program, U.S. to Switzerland via *Early Bird* communication satellite
1967	Estimated over 56 million homes in U.S. have television sets
1970	Low cost glass fibers produced for fiber optic cables
1975	Silverless photo film produced
1986	Filmless photography—all electronic camera used

Visual Communication

A newspaper represents a printed or visual medium of communication. The information source may be an artist, reporter-writer, or cartoonist. The message is transmitted by the printing press on which the newspaper is printed. What would be the carrier or channel? The paper on which the news is printed. In printed communication, the eyes are the receiver of the visual message. The brain then decodes the printed message.

Smeared printing is an example of visual interference. If the pictures in a book were printed poorly, one might not be able to identify the picture. This also would be interference. Another example is a comic strip that has the colors printed out of the line drawing.

People have used their senses of sight and hearing for thousands of years. When these senses were not enough, new methods of sending messages were developed.

History of Communication

Early Methods

Early people painted pictures on the walls of their caves. There was no alphabet. Cave dwellers drew pictures of what they saw. The Egyptians were among the first to develop an alphabet, using symbols called *hieroglyphics*. Some symbols were carved in stone and have been translated into modern languages.

The Greeks and Romans developed alphabets that are used today. These symbols made possible the communication between people. History was recorded, messages were sent, and books were written. Every book had to be hand copied. As a result, progress in communication was slow. Imagine what it would be like if there were only one copy of this book. Classmates would have to copy to make their own books.

Mass Printing

Books first were printed by a method of stamping. Each sheet of paper was stamped with a carved block of wood coated with ink. This began in Europe in the early 1400s. The words and pictures for each page had to be carved of wood. This process was long and difficult. Only the wealthy could afford the first books printed.

Movable metal type was developed in Europe and Asia. In Europe Johann Gutenberg was recognized for his work in printing a Bible. This was the first large book printed with movable type. Each letter was cast separately. This made it possible to use the

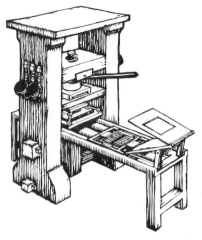

A printing press in Gutenberg's time.

letters as many times as needed. Gutenberg's Bible of more than 1,000 pages was printed one page at a time. The Bible was printed around the mid 1400s.

Printing began in colonial America in 1639. A printing press was used to print the colonists' first books. A notable printer of the time was Benjamin Franklin. He made many contributions to the letterpress printing industry.

In the late 1700s a new method of printing was introduced. Alois Senefelder invented the *lithographic* process. *Litho* refers to the stone that was used to print. The smooth stone was etched with a chemical and coated with gum to prevent ink from sticking on the unetched areas. The image was then transferred to paper. This printing resulted in a reverse or mirror image. Senefelder printed charts, prayer books, portraits, textbooks, and landscapes.

A lithographic rotary press was developed around the 1800s, creating interest among printers. Two young American lithographic printers became very successful in the 1850s. They printed unique, recognizable scenes of American development. Their works have sold millions of copies. James Ives and Nathanial Currier became known as Currier and Ives, who printed over 7,000 different scenes.

In 1905, Ira Rubel discovered the effect of printing by offset. The term *offset* is used to describe printing from an image setoff from the original, reversed print. As described earlier, lithographic printing resulted from a reverse image. Offset lithography does not require a reverse print-making process. Printers could work from a correct image rather than a reverse one. The offset process also improved the image on the paper. This was accomplished by putting ink on the correct-reading plate and transferring it to a blanket cylinder. This cylinder is covered with a **rubber material. A study of the drawing on page 246 will help you** see how the plate starts as a positive image. It is then transferred to a rubber blanket. Thus, the rubber blanket prints the image correct on paper. The process is called offset because the printing is done by an image that is set off by the plate.

Rubel developed a press to use this process. The press was called an offset press. Today the offset press is very popular. It grew in popularity with the photographic process. The reasons are linked to the way printing plates are made. Around 1915 offset plates were being made from photo negatives. This made it possible to print photographs in color. The photo process simplified the printing process as compared to the letter press process. Many letterpress print shops began adding offset equipment.

Photography

Early photography used metal and glass plates with light sensitive materials on them. These were used to make the first photographs. Matthew Brady, the Civil War photographer, used a wet type of negative developed by a Frenchman, Louis Daguerre. Brady's photographs were detailed and immortalized his subjects as well as himself.

George Eastman developed the first flexible negative material. This eliminated glass from the negative. It also marked the beginning of the Eastman Kodak Company.

The simplest camera is the pinhole camera. It is easily made by using an oatmeal can or box with overlapping lid. A pinhole camera can also be made from a film cartridge. A hole in the front allows light to strike the film or paper sealed inside. A piece of black tape to seal the opening acts as the shutter (see the pinhole camera activity in the Activities Manual).

This camera has some unusual features one should consider. By moving the film closer to the opening a wide-angle lens is created. By moving the film away from the opening, a telephoto lens is created. One limiting feature is that the exposure time is usually long. This means one must support the camera on a solid object like a chair or table while taking the photograph.

The pinhole camera can provide many interesting photographs. These photos will appear as reverse images on film and must be printed on paper to become a positive photograph. A pinhole camera does not require a lens. The lens is responsible for inverting the image.

Early cameras developed by Eastman for public use had to be sent back to the manufacturer for developing and printing of the film inside the camera. For $25.00 the photographer got a camera and 100 exposures. Many types of cameras have evolved from this early camera.

Modern cameras are identified by their lens arrangement. Cameras that use a mirror are usually called *reflex*. The mirror is used to reflect the image for viewing. There are two common types, single lens and twin lens. When using a camera such as these, there are three adjustments that have to be made before taking a picture.

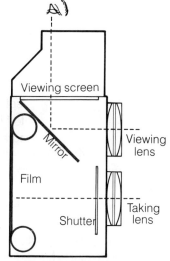

Twin Lens Reflex Camera

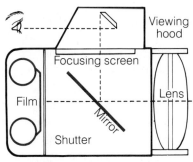

Single Lens Reflex Camera

1. Focus of the lens for sharpness of the object.

2. *f* stop adjustment to control the amount of light striking the film. (Ranges from *f*:1 to *f*:32.)

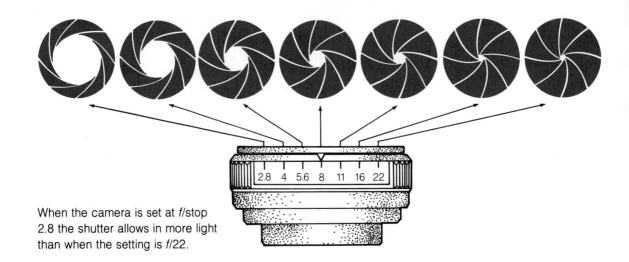

When the camera is set at f/stop 2.8 the shutter allows in more light than when the setting is f/22.

3. Adjustment to control the time the light strikes the film. Commonly used shutter speeds range from one second to 1/1000 (one thousandth) of a second.

SHUTTER SPEEDS

1 Sec.	1/8	1/15	1/30	1/60	1/125	1/250	1/500	1/1000

Automatic cameras will perform all the adjustments named above. An automatic camera can be very simple, such as an Instamatic. Some automatic cameras can be very expensive. They have electronic circuits that measure light and make exposures.

One type of camera can produce a finished color photo in sixty seconds. The color image develops in room light. This is different from all other photography, which must develop in darkness. The new film is the secret to these types of cameras.

The film contains multilayers of chemical dyes (eighteen to twenty layers) and a developer. The materials are first exposed to light. Light, itself, is composed of all colors. For example, when one sees a red rose, the rose is absorbing all colors in the light except red. When light strikes the film it causes a chemical change to occur in the color dye. The red rose would cause the red dye to be exposed, making that portion of the photo red. A chemical in the photo acts as a developer to cause the dye to change colors. This chemical is released when it passes between two steel rollers in the camera.

The instant-picture camera has a great advantage. It produces a finished picture in only sixty seconds. One does not have to wait several days for the film to be developed. However, it is more expensive than other films.

A form of photography called *holography* produces a three dimensional image much like the actual object. This effect requires a laser beam instead of a camera and film.

Photography and Printing

The development of the camera and photography in the mid-1800s gave Matthew Brady a tool to record history. He photographed the events of the Civil War from 1861 to 1865. Later in the 1870s photography changed lithography. Photography made the printing of photos simple. Lithographic printing became offset lithography.

An offset press was built after the turn of the century. The offset press of today is much faster. It can print and register many colors in accurate position with other colors. A modern offset press is composed of many systems. The basic system includes paper feeding, inking, moistening, impression, and delivery. Each one has a special job.

The paper feed can be simple or complex, depending on the paper it must feed at the speed of the press. The paper feed part of a press would include a vacuum system to pick up the sheet, a double sheet preventor, and register board.

To understand the process better, let us follow some paper through a press. Paper behaves in ways that a press must overcome if it is to run at speeds to ten thousands of sheets per hour. Paper sticks to itself and must be blown apart by blowers. So as the suction feet pick up the paper, the blower separates them.

The paper is then fed into rollers that are driven at the speed of the press. These rollers carry the paper to the register board. Here the jogger pushes the paper to one stationary side which becomes the common point for determining where the image will be printed. A set of grippers opens and grasps the paper. The paper is pulled through the printing cylinders and then from the press.

The inking and moistening systems are very carefully balanced. Too much of either one will cause problems ranging from no image to covering the paper with solid ink. Once the balance is correct, printing will occur properly. The ink and water mixture is applied to a plate. The ink will stick to the areas which light has

Offset Printing

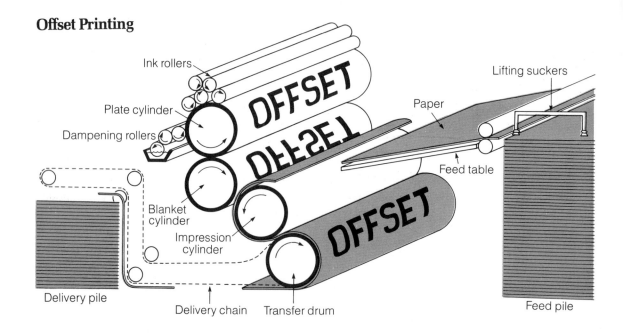

struck. The water mixture is applied to the entire plate. Since the ink has an oil base and oil and water do not mix, the water mixture does not stick to the ink coated areas of the plate.

The plate will now come in contact with a rubber mat called a blanket. The ink will then transfer much like the process of a rubber stamp and ink pad.

The paper then is fed between the rubber blanket roller and the steel impression roller. The ink transfers from the rubber blanket to the paper. The printed paper then is delivered to a jogger for future handling.

The image from the metal impression plate is transferred to the rubber blanket cylinder.

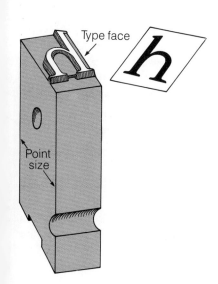

A pressman checks a sheet printed on this large web press.

A press used to print newspapers is sometimes two stories tall and has paper fed in large rolls. Other types of offset presses print two, three, or four colors at the same time. Think of the many products that are printed. Specially designed presses may be required to print many of them.

Letterpress Printing
Letterpress is a term to describe printing from a raised surface. The letters used in printing are called type. A piece of type is a mirror image of the actual letter. Reverse type is arranged on the bed of a press. A large impression cylinder is used to press the paper firmly against the ink-coated type.

The type of today is made of metal, and instead of one letter set at a time, whole lines are cast by typesetting machines, often called linotypes. A more recent development is photo-composition which uses no metal type, but produces the image on film.

A press can print much faster if the plate is cast into a circular shape. This allows the press to print on each rotation of the press. This method is used to print newspapers. Many larger city newspapers are switching from letterpress to offset although much of the letterpress equipment is quite expensive to replace, so it must be used for many years, even though newer machines may be available.

Communication

Electronic Message

Electricity was discovered by Benjamin Franklin. At the time, he didn't perceive its many potential uses. Samuel F. B. Morse is credited with the first electrical device for communication. Insulated copper wire wrapped around a metal rod will become a magnet when it is connected to electricity. The metal rod will draw other metal objects into contact. When this occurs a clicking sound is made. When electricity is disconnected, the object springs back.

From this discovery, Morse developed a code of short and long sounds, known as dots and dashes. These sounds in combinations are used to communicate. Samuel Morse sent a dots and dashes message in 1844. This code is an international language and takes only a short time to complete. The first single-line telegraph was constructed between Washington, D.C. and Baltimore, Maryland.

Electrical experiments were done to find a method of sending a human voice by wire. In 1876 Alexander Graham Bell succeeded in sending a spoken message over a wire. Transatlantic telephone messages began in 1956. Until then, the cables could only carry telegraph messages. The telephone provided communication between the United States and Europe. The important fact about the phone was that it was, and still is, an almost instant form of communication.

Radio signals were developed in 1895 by Guglielmo Marconi. He used a long wire and an antenna to transmit the signal. Radio signals were used like a telegraph, except that no wires were connected between cities. (It was called a *wireless*.) Voice transmission was developed soon after. In 1906, Dr. Lee DeForest developed a vacuum tube to increase the weak radio signals. In 1920, the first commercial radio station went on the air.

Television was introduced officially in 1941. In the late 1940s, television sets for the home were developed. It was not very popular in the beginning because of a lack of programs of interest to the public. People didn't seem to be interested in seeing a newscaster read the news on television. Today very few American homes are without at least one television set.

Advancements in electronics have expanded the use of radio and television. The transistor was invented in 1947 by two scientists at the Bell Laboratories. Its development made modern lightweight communication devices possible. Today satellites circle the earth, keeping us in constant contact with the world. One of the first satellites was Telstar in 1962. This satellite was placed in orbit over the equator.

Development of Telecommunication Media

Telecommunication

Telecommunication means communication by use of electronics. Examples are the telegraph, radio, television and computers.

Telephone. In March, 1876, Alexander Graham Bell and his assistant, Thomas Watson, connected wires to a diaphragm, which was backed by an electromagnet. Bell believed his voice would move a thin metal disk toward the magnet and then away. This motion would cause electricity to flow through the wires connected to a diaphragm. This instrument was connected by wires to another similar instrument. These wires carried the current, which caused the diaphragm to move identically with the other instrument. His changing voice caused changes in the electrical current. This movement of the diaphragm caused the air around it to vibrate, creating a sound. Thus, the first spoken message was sent over the wires.

The telephone gave people the ability to communicate over great distances. In more than 5,000 years of history, this was the first time a message was sent over any great distance in a short time. This message was sent at the speed of light, 300,192 kilometers (186,000 miles) per second.

The telephone system of today has developed into a system of worldwide communication. All telephones have a number to direct a person to the right phone. The United States and Canada are divided into more than 100 area codes. Phone calls by the millions are made every hour.

Local calls are carried over cables of wire. These cables are large, heavy, and expensive. In cities, these cables often are buried underground. They are connected to a series of switches located at a central office for that community. In early telephone systems, connections were made by a telephone operator. The dial system changed all this, and today computers connect more than 550,000 phone conversations every hour.

Messages are now being sent through glass cables. This is called *fiber optic* communication or simply, "fiber optics." Optical communication has always been a part of our lives. Hand signals make use of light but are of little use at night. Traffic lights and signal lamps on automobiles are also optical. Signal lamps are used on ships and in control towers at airports.

Sophisticated optical communication is not new. In 1880 Alexander Graham Bell invented what he called a "photophone." Sunlight was reflected from a mirror. Sounds of a person's voice made the mirror move slightly (modulate), and this light was picked

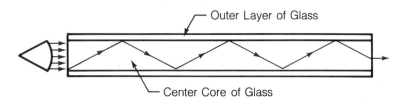

The outer layer of glass causes reflection of light within the fiber.

Different types of fiber optic cables.

Underground fiber optic cables may eventually replace overhead telephone wire. This installation is at Cape Canaveral, Florida.

up by the receiver which converted the signal to electricity. A telephone receiver was connected so the electrical signal could be heard. In this way a message was transmitted optically, or over a light beam. Though the photophone worked well, it was never accepted.

Practical fiber optic communication was not possible until 1970 when low-cost glass fibers were produced. A good light source—the laser—already existed. Invented in 1960, the laser was already being used in many ways.

Messages could be sent by modulating (varying) the laser beam. The lasers at this time were very limited: they could only transmit in a straight line; the light beams could not be bent without losing

the message; weather conditions affected the laser beams; clouds would block the beam. The laser beam was also dangerous— a person looking into the beam could be blinded. There was a need for some way to control the laser beam. Fiber optics was the answer.

Fiber optic cables are made up of many tiny glass strands or fibers. Some of these fibers are smaller than a human hair! Light is put into one end of the fiber or cable. It is reflected within each fiber and comes out the other end. When carefully made and coated almost no light is lost even through very long cables. The cables can even be bent around sharp corners without loss of the light.

When made into cables, the fibers are very rugged and can stand a lot of wear and tear. The fiber optic cables are not affected by corrosion, radiation, magnetic fields, and electrical lines and generators which interfere with messages going through copper cables.

Fiber optic cables have the added advantage of being much smaller and lighter than metal cables. A standard wire telephone cable contains 900 pairs of twisted wire. It is 70 mm in diameter and has a capacity of 21,600 calls. One fiber optic cable made for telephone use contains 144 fibers. It is only 12.7 mm in diameter and has a capacity of 96,768 calls. The fiber optic cable is 30 times smaller but can handle more than four times the number of calls.

Fiber optic cables are used for many purposes. They are used to transmit light into hard-to-reach places; even inside the human body. Doctors use special scopes that allow them to look inside the body or to provide light for delicate surgery.

Fiber optic cables are used to light parts of the dashboard in some automobiles. A single lamp provides the light. Several fiber optic cables transmit the light to various parts of the display on the dash. Robots or guided missiles can be controlled through fiber optic cables. If a television camera is placed in the robot, the operator can see where the robot is going and control its movements very accurately. Two-way communication is possible through the cable so the operator can both see and send instructions through it.

Fiber optic cables are also being used in sensors for hard-to-reach or dangerous places. Temperature, pressure, liquid levels and even readings of instruments in radioactive areas can be transmitted through the cables. In many cases this could not be done safely or as efficiently with metal cables.

Even though telephone companies are developing microwave transmission of phone calls, repair service is still necessary on the regular lines.

Telephone companies have developed many new ways of sending telephone calls over great distances. One method used in some areas is microwave transmission. This system transmits phone calls through the airways, in much the same way as radio is transmitted. Microwave is used because it is cheaper than stringing more copper wires. Microwave towers are used to transmit and receive messages.

Communication satellites have been launched into orbit around the earth. These satellites are used by the telephone companies to relay calls around the earth. Television networks have invested millions of dollars in satellites to speed news reports from all parts of the earth. You can watch televised reports of events as they happen from nearly any place. The American Broadcasting Company broadcast the 1976 Olympics from Japan. Millions of Americans saw the events as they occurred. This was the first continuous live transpacific television broadcast. The newly orbited SYMCOM 3 satellite was used to relay the broadcast signals.

SYNCOM 3 was put in a "geostationary orbit" at a particular height above the earth (about 22,500 miles or 36,000 km). Its orbiting speed was very precise (about 6,879 mph). When in the proper orbit, the satellite remained over one spot on the earth, appearing to be stationary.

A geostationary satellite is visible from slightly less than one-half of the earth's surface. Three such satellites at a high altitude can cover all of the earth (except the extreme North and South poles).

Early communication satellites were very limited in their use. They could only reflect radio signals from one point on earth to another point. Television signals could not be transmitted. Modern satellites are very complex machines. They are often called "repeaters in space" because they amplify the signals before returning them to earth.

Modern communication satellites have antennas to receive signals and to return them back to earth. The signal that is received by the satellite is very weak. It loses most of its strength on the long trip from earth. Equipment in the satellite amplifies the signal. So that this stronger outgoing signal does not interfere with the weak, incoming signal, transponders are used to change the frequency of the two signals. This eliminates the interference.

Communication satellites are actually machines made up of many very complex sub-systems. Power is provided by an array of solar cells. Batteries provide back-up power for times when solar energy is not available. The satellite must be kept in perfect posi-

Geostationary Satellite

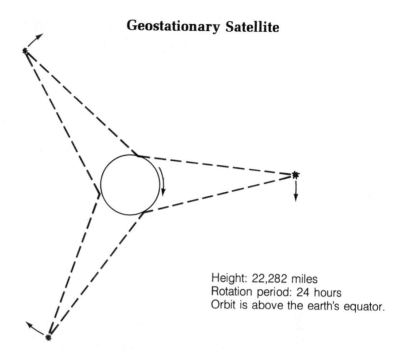

Height: 22,282 miles
Rotation period: 24 hours
Orbit is above the earth's equator.

tion. A system of stabalization rockets is used for this purpose. A temperature control system keeps the satellite at the proper working temperature. Many other sub-systems are required to keep the satellite working properly during the years of its expected life.

The tracking station in Somis, California tracks and controls two communications satellites.

Communication

253

A satellite called the ATS-6 was launched May 30, 1974. It was a National Aeronautics and Space Administration project designed to rebroadcast programs to remote areas of the world. Alaska and other mountainous areas where local television reception is very poor have been test areas. The ATS-6 satellite made it possible for people in remote areas to receive television signals from space.

The ATS-6 had a nine-meter diameter antenna. The antenna was made of metal mesh. It was folded into a tight package for launching. When in orbit, the mesh was opened to reflect television signals. The signals are sent to the satellite, which reflects them to antennas on earth. In this way the television signals can go over mountains and around the earth.

This new communications satellite is twice as big as the old model and four times as powerful. It is nearly two stories tall, weighs about 2,550 pounds. This satellite provides voice, TV and high speed data services to Indonesia, the Phillippines, Malaysia, Singapore and Papua New Guinea.

Radio. Each radio station is assigned an operating frequency. This frequency makes it possible to have more than one station broadcasting at one time. The total radio frequency range is from 10,000 cycles per second, written 10 KHZ, to 300 million cycles per second, written 3,000,000 KHZ. Each station is allowed to operate a very large amplifier at the assigned frequency. Since the radio frequencies start 10,000 KHZ above the hearing range of the human ear, this large signal cannot be heard. The human ear can hear sounds from 20 cycles per second to 20,000 cycles per second, 20 KHZ. HZ is an abbreviation to "Hertz" which means cycles per second. The name Hertz is an international term to honor the man who made great contributions to the field of electronics.

The power of a 5000 watt transmitter could light five hundred 100-watt light bulbs. This transmitter sends its power to an antenna atop a tall tower. The energy is then radiated into the air. The signal is reflected off the ionosphere to earth and received by radios. The signals we hear, music or voice, are traveling at the speed of light.

Suppose you and two friends with walkie-talkies try an experiment to find out which is faster, radio waves or sound waves. A good place to try this experiment would be a football field.

Position one friend (A) with a walkie-talkie at one of the goal lines. The friend (B) without a walkie-talkie and you with your walkie-talkie should walk to the other goal. Place a finger in one ear so you cannot hear friend (A) at the opposite goal. Now test your radio. Have friend (A) shout out some command that he could see you do. If you heard him, raise your right hand. Both you and friend (B) should respond. Now begin the experiment with friend (A) pressing the talk button on his walkie-talkie and shouting the command. Friend (A) will be the judge to see who raises his hand first. You or your friend (B)?

In theory, sounds travel at 350 meters (385 yards) per second in air, while radio waves travel at 300,192 kilometers (186,000 miles) per second. Any voice could be transmitted around the world by a radio transmitter. Radio waves are all around us. They come into our homes, schools, cars, etc. To see how many radio stations are sending signals to your home, turn on an AM radio and set it on the left end of the dial, 5.4 KHZ. Slowly turn the dial marking the first station by writing down the dial reading. If you have two stations close together write a 2 beside the first dial reading. A good receiver on a rainy day could receive thirty to forty clear stations. More stations can be heard at night because

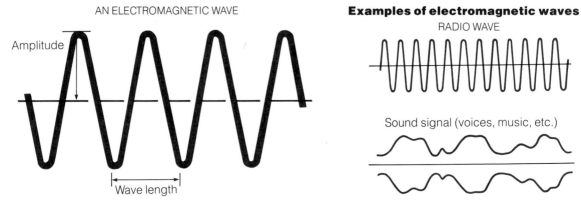

Radio Signals and Sound Signals as Electromagnetic Waves

the earth's atmosphere changes to favor radio waves. Other factors that affect radio reception are weather, frequency of the signal, mountains, and buildings.

Radios are either AM, amplitude modulation, or FM, frequency modulation. These terms describe the methods of modulating (regulating) a signal. Radio waves cannot be heard. It is the music and voice sound waves that we hear. The two signals, radio frequency and audio, are combined to form one signal. The modulator does this by taking the voice and music sound waves and controlling either the amplitude or frequency of the radio wave.

FM radio has less static than AM radio because a static wave does not affect FM reception. AM radio waves reflect off the

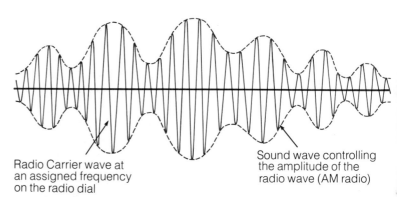

Amplitude Modulation

atmosphere and static is received during this process. FM radio waves tend to travel in a straight line. This limits their reception to 161 kilometers (100 miles). AM stations are usually less powerful than FM; however, AM will travel farther distances because these waves bounce off the atmosphere.

Some local AM radio stations are limited to daytime operation and power to cover a 25 to 45 kilometer (15 to 30 mile) area. This limit allows each area to have its own radio station. No matter where one visits in the United States, a radio program can be heard with little interference from any other radio station that may have the same place on the dial.

AM radio transmission is based on the use of the *ionosphere*. The ionosphere is a layer of charged particles around the earth. The layer changes during a 24-hour period. This explains why at night, when the charged particles are closer together, AM radio reception is better. More radio waves are reflected at night, thereby increasing radio reception.

Citizen Band radio is an example of an AM radio. The term CB became very popular in 1975-76. The gasoline shortage during the summer of 1975 caused truck drivers to seek a method of finding gas supplies for their trucks. The CB radio helped solve their problems of gas supplies and slower speed limits. Truckers could communicate from 6 to 16 kilometers (four to ten miles) up and down the highway, warning of radar patrols and telling where fuel was available. Since 1975, CBs have grown at the rate of 20 to 30 million sets per year.

The CB radio is a combination transmitter and receiver. It operates on specific frequencies assigned by the Federal Com-

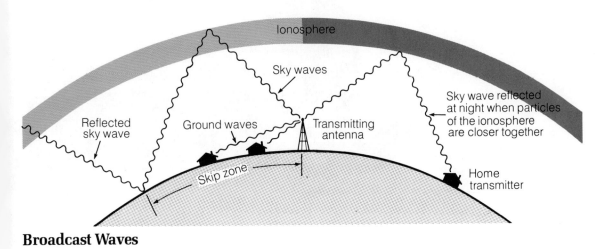

Broadcast Waves

munication Commission (FCC). The frequency range is 26.97 to 22.27 million Hertz divided into 40 channels. The transmitter is crystal controlled to prevent interference from one channel being heard on another channel.

It is very important that CB operators have an FCC license before operating their radio. This license may be obtained by filling out an application form and sending it to FCC headquarters. There are other requirements such as a minimum age of eighteen, and U.S. citizenship. Once one has a license, one is given a call number which is used to identify oneself on the air. The proper transmission on the air will include one's call numbers and a nickname or so called "handle." KJE 5268 is a call number and the "Gaggon Dragon" is a handle that could be used by a CBer for identification.

Mobile CBs operate in cars, boats, or bikes. Base units are found in the home with antennas up to 18 meters (60 feet) high. CB radios are limited to a power of four watts input. The watt is a unit of measurement of electrical power. Four watts is not much compared to AM broadcasting stations operating at 5000 watts. However, the FCC has very good reasons for limiting the power. One reason is that a CBer operating on higher wattage could transmit farther than he could receive. This would interrupt other CBers' transmissions for miles.

The CB transmitter can transmit on any of the forty chanels. The receiver also will receive on any of the forty channels. When

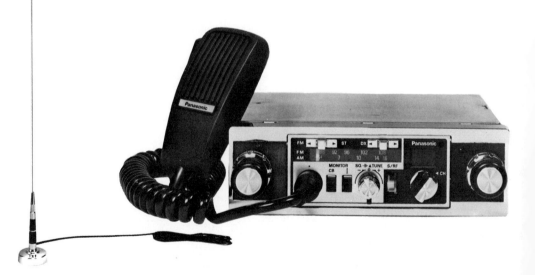

CB radio and antenna

a channel is selected, the CB's tuner is locked on one frequency. This is the frequency of the carrier wave. The carrier wave is not heard. To transmit a message, one must modulate the carrier wave at the rate of one's voice. This is done when the transmitter key on the microphone is pressed.

One switches a radio from transmitting to receiving at the press of the microphone switch. The carrier wave modulated by the voice is sent to the antenna. The cable linking the transmitter and the antenna is called the transmission line. It must be of proper size and length for maximum distance. Antennas are available in many shapes to fit any need. Mobile and base antennas are very different. Mobile antennas are usually smaller.

CB transmissions are in the form of ground waves with some waves reflecting off the ionosphere. Reception and transmission are affected by weather conditions. Under certain weather conditions, a radio wave can reflect to earth hundreds of miles away. CBers in Georgia might be heard in Virginia. This condition is called *skip*. The difference between where the ground wave stops and where skip is heard, is called the *skip zone*.

Television Broadcasting. Television was developed about the same time as radio. A Russian scientist, John Logie Baild, invented an electromechanical television. There were a few thousand of these sets in England, but World War II stopped further television progress. Television today has been used on the moon to send live pictures to earth.

Television is described by its name *tele*, meaning tell or talk, and vision, meaning see. The television set is a receiver of two radio signals sent from broadcast stations, usually within 100 kilometers (65 miles) of one's home. The reasons for this distance are many. First, television signals travel by line of site, which means mountains and tall buildings interfere with reception. Secondly, each community throughout the United States has its own likes and dislikes about what it wants to see. Would you be interested in the weather report or news of a community 300 kilometers (180 miles) away?

The broadcast station is authorized to operate on only one channel from channels 2 through 13 on VHF (very high frequency) or channels 14 through 84 on UHF (ultra high frequency).

Each channel represents a separate frequency set by the FCC. This is just like the stations on a radio dial. This prevents more than one station from operating on the same channel.

The first part or circuit of the television to be discussed is the tuner. One might call it the television's brain, since it controls which channel is shown on the screen. The process of taking a very weak signal from the antenna and converting it to a stronger, lower frequency is the other job of the tuner.

The tuner gets its power to operate from the television's direct current power supply. This power supply has the big job of supplying many different voltages. This could be 20 volts (d.c.) of electrical pressure to the 25,000 volts needed to operate the picture tube.

The picture tube or cathode ray tube (CRT) provides the picture. This is accomplished by showing thirty still pictures in one second. This appears to the human eyes as normal motion. The CRT is composed of an electron gun and a phosphorous coating on the face of the tube.

The electrons are directed from left to right and top to bottom, much like a toy Etch-a-Sketch. Just as one draws pictures with an Etch-a-Sketch, a picture is drawn on the fact of the CRT with a beam of electrons. Electrons are the very small negative particles of an atom.

Color picture tubes have three electron guns to give red, blue, and green colors. The color does not come from the electrons, but from the different phosphorous colors on the face of the CRT.

The audio portion of the television is an FM radio receiver. The audio signal is transmitted along with the video or picture information. This video signal is amplified enough to power the CRT and produce moving pictures on the screen.

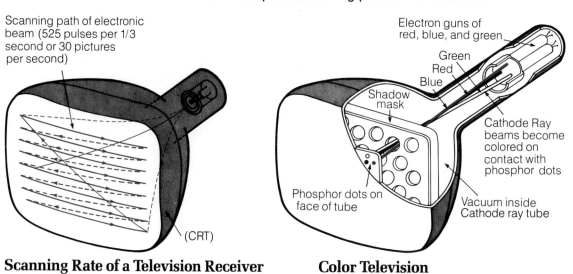

Scanning Rate of a Television Receiver

Color Television

Televisions have a set of controls that the viewer can adjust and another set that only a trained serviceperson should adjust. The brightness and contrast are two controls that effect the operation of the picture tube. Some viewers keep the brightness set too high, washing out picture detail. Contrast helps the viewer separate the difference between white, gray, and black. These important adjustments concern the black and white portion of the picture. A good black and white picture comes first, then color adjustments.

Color adjustments are on sets for those who want a particular color level or skin tone. For those who do not want to adjust their own, preset adjustments are included in an automatic color set.

Controls on the back of the television are usually for the serviceperson. The only exception would be the vertical linearity, vertical hold or horizontal hold. One should not adjust color controls on the back.

The television has become the home entertainment center. Television games and closed-circuit video tapes are providing more uses for television sets. Cable television is providing perfect reception using phone lines instead of an outside antenna. The cable also increases program selection using all channels. Educational and public television also give the viewer a choice of programming.

What new uses can you imagine for the family television? It could become a computer terminal, linking the family with local banks, grocery stores, or library. An attachment for a television that will play a video disc of a prerecorded program is now available. A new system of tuning called V.I.R. is being used by General Electric Corporation. The color tuning on the television will be controlled by the station. This will give true color as transmitted. A new feature being added to cable vision allows the viewer to answer questions and vote for contestants on programs.

Message Generation—Mass Media

To generate a clear message one must understand the person or persons who will receive the message and aim the message to that person. For example, the message in a comic book usually is for a young reader. Another example would be the evening news for adults.

Advertisements. Advertisements are examples of a selling message. Companies like Goodyear, Coca Cola, and Ford spend large amounts of money to develop advertisements about their

A control room of a television studio.

products. People who write slogans for these companies must be very creative. They are generating original ideas about a product. The ideas or slogans will be posted on billboards, seen on television, and heard on the radio. Can you think of some slogans about a motorcycle or a car? Slogans can become very popular. Play a game: Say a popular slogan and see if your classmates can guess the product it identifies. For example, "You deserve a break today" is a slogan for McDonald's food.

Writing a song or a book also is generating a message. In each case the author considers the people who are going to hear or read the message. A message can take many forms and be verbalized, visualized, and/or touched. Some messages stay in mind because they are heard over and over again.

Electronic Television Production. How is a television show produced? The first step is to survey the current programs or review the kind of programs that are currently on the air. Then a decision is made if a need exists for another program. If so, what type is needed; comedy, drama, game, or variety show. After this decision has been made, a word-for-word dialogue is written. The dialogue or script is developed by writing teams. A comedy show may have a large staff of writers who write a new show every week.

Once a script has been written, the *actors* and *actresses* must learn their parts. Memorizing is the first step. The interpretation of the words and events, which gives them meaning, comes next.

Finally, the actors and actresses must perform their parts before a camera and/or audience.

The *producer* is the person who makes everything happen. This person works with every one involved: casting, acting crews, director, sound people. The producer also works to get necessary equipment, materials, costumes, and personnel. Producing a television show often starts with an idea for the theme of the show. Think about the different shows that are on television. What new theme do you think would make a good television show?

The *director* works with actors and actresses to help them interpret their roles in the show. The director sees the acting crew as we see them. Directors help each acting person to give the best performance possible.

Television shows usually are taped and broadcast later. However, some shows are aired live, such as football games, parades, news programs, or daily talk shows. A director works behind the cameras in the control room to record each act. The final tape will have each portion of the performance recorded together with no time lost for costume or scene changes.

The director also must direct each camera to the right position or angle. There are usually three cameras: one fixed on either side of the stage and one on the back of a roving camera operator. Getting the best possible camera angle at the right time is very difficult. Often several cameras are used, though we only see the signal from one camera on our screen. This gives a variety of angles and creates special effects. Several cameras are used to cover all the action of a football game. The director decides what we see on our set. He monitors all cameras and switches from one camera to another. Directing is like a performance in that it requires the director's full attention. Some parts can be practiced but other decisions have to be made during the taping or airing of the show.

Message Reception—Mass Media

The form of communication most often regarded as mass communication is television. An average of two television sets per household exists in the United States. Television has linked the nations of the world. This technology has brought millions of viewers entertainment and information. Radio as a medium also spans the globe and distant planets like Mars, 368 million kilometers (230 million miles) away. Radio programs are both entertaining and informative. Magazines, newspapers, and books make up the third mass medium. Libraries store the materials as a record of our achievements, history, and hopes.

Machine Message. While tuning a radio, driving a car, or typing, one is communicating thoughts to a machine. Communication with machines is a big part of our lives. A better understanding of this process will lead to improved communication. As one grows in knowledge about a machine's operation, so will one's ability to operate the machine.

Message Processing. Processing a message means changing the form. For example, the telephone company processes phone messages each day. A voice is changed into electrical impulses. These electrical impulses are amplified and switched.

Amplification is done by electronic circuits like the phonograph amplifier. The volume and tone of a phone call can be changed just as one changes the volume and tone on a phonograph.

Switching is the most complex process. The phone company must have miles and miles of wire to almost every home in the country. Each home has a unique phone number. This phone number, when dialed from anywhere in the United States, will ring only in that house. The phone company's central offices house a maze of electronic switches. The switches are automatically controlled by the code sent by someone dialing a number. When the circle digit is dialed, long-distance circuits are opened. The circuits are now ready to receive an area code which identifies a part of the country or of a state.

The phone company processes many types of messages other than phone calls. High-speed computers communicate large amounts of data from city to city by phone lines. Television and radio programs are sent over phone lines from New York to cities across the country. Otherwise, these programs would have to be recorded and mailed.

Message Reception. Electronic signals transmitted from television and radio stations are received by antennas. Antennas are

Sound Wave Being Amplified

designed on the basis of the length of a standing wave of the frequency being transmitted. This simply means each frequency has a wave length that is used to determine the length of the antenna. Therefore, each frequency has a different length antenna. An FM radio has an antenna length of about one meter (three to four feet).

Television antennas are shorter than some FM antennas. The higher frequency television channels have shorter antennas.

Telephone companies also use antennas for receiving microwave signals. This antenna looks like a deep dish mounted on a tower. The receiving antenna is directed toward the transmission antenna. This is usually a straight-line reception pattern.

When a person receives a message, it is directed at one of our five senses: taste, sight, touch, smell, or hearing. Messages have more impact when we see and hear the message. Written information does require interpretation through reading. Messages from our senses are translated instantly by our brain. We rely on our senses for learning, earning a living, and communicating with others.

Impact of Communication

How has the communication explosion affected our lives? A family would experience this impact through the stockmarket report in the daily newspaper, listening to a new recording on the radio, clipping a recipe from a magazine, or watching the day's events on a television news program. Other effects are buying a new product that was advertised on television, checking out a book on gardening from the library, or getting a phone call from a fellow worker with the results of yesterday's work. These examples point out the many ways communication media affect our lives.

Communication has an impact on our government. Our government could not operate today without mass media. Communication serves all levels of government: local, state, and national. Congress has all the speeches and bills printed daily in a document called the Congressional Record. State governments print and distribute documents to provide services for us. Our government leaders keep us informed through the media.

Our neighbor, Canada, broadcasts sessions of its Parliament. These broadcasts place government leaders under the scrutiny of the television viewers. Canadian citizens are becoming better informed.

Interactive Video

On a trip through a department store you come across a display for a new camera. Included in the display is what appears to be a television set with a show playing on it. The features of the camera are being explained by one of the world's best photographers. But this is no ordinary television program. You can interact with it. You touch the screen at a certain spot and the program changes. Your answers to questions tell the computer in the display how much you know about cameras. The computer then selects information you need.

If you decide to buy the camera, you can insert a credit card in a card reader on the display. In several days your camera will be delivered to your home. You have bought a new camera and never talked to a salesman. You did not have to be in a camera store. The display could have been set up anywhere.

How is this possible? An interactive video system was used in the display. A video disc player was connected to a computer. Still pictures and motion pictures were stored on video discs along with printed information. The complete story of the camera was put on the disc. If you indicated you did not know something about the camera, the computer selected that information to be shown. Your answer to a question told the computer what information you needed. It was selected from the millions of bits of data on the disc and shown on the monitor.

Interactive video is being used in many different ways. Point-of-sale (where the item is being sold) displays such as the one described above are used in department stores. Similar displays are used to give tourists information about nearby hotels, restaurants and scenic spots. A display in the Statute of Liberty gives information and shows photos of early immigration into the United States.

Interactive video is becoming a valuable educational tool. Business and industry are using the systems in on-the-job training. Auto mechanics can learn the latest repair techniques from the country's top mechanics — without leaving their own shops. They can even practice what they are learning as they watch the monitor. People learning to give CPR can practice on a computerized interactive dummy. The interactive system tells the student how well the techniques are being done. Students can be given a grade indicating how well they learned the techniques.

Audio and Video Discs

Compact disc (CD) players are very popular. They are rapidly replacing audio tapes and records. This is due, in part, to the very clear sound produced by the system and the fact that playing the disc does not wear it out.

A CD is made by molding plastic into a 4.7 inch diameter circle. Grooves and fine pits which produce the sound are impressed into one side of the disc. This side is coated with a shiny aluminum film and a thin protective layer of plastic.

A laser in the player bounces light off the pits in the disc. This is picked up by an optical sensor. The laser tracks around the grooves similar to the tone arm in a record player. The light signal picked up by the sensor is sent through a microcomputer to convert it into analog form. This is amplified and played through speakers. Very clear sound without any background noise is the result.

The compact disc has great potential for future uses. One 14-inch disc can contain 50,000 pages of text. 1,500 floppy discs would be needed to hold the same amount of information. As many as 50,000 pictures can be stored on a 14-inch disc. It has been estimated that the entire 21 volumes of the Grolier Encyclopedia can be stored on just one-sixth of a fourteen-inch disc. Specific words can be referenced in seconds. Imagine how easy it would be to do research when topics can be found that fast!

Though the CD was not marketed until 1982, it is a very popular piece of home entertainment equipment. In 1986, CD players were

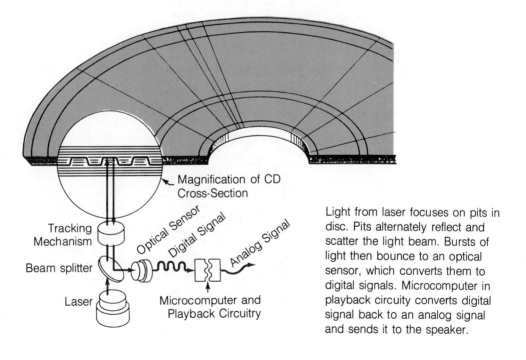

Light from laser focuses on pits in disc. Pits alternately reflect and scatter the light beam. Bursts of light then bounce to an optical sensor, which converts them to digital signals. Microcomputer in playback circuity converts digital signal back to an analog signal and sends it to the speaker.

Communication

the fastest selling piece of hardware in the history of home electronics.

In the Museum of Immigration in the Statue of Liberty in New York, video discs are used to store thousands of photographs. The photos were collected from people from across the United States to show some of the immigrants who came to this country. Visitors can stand at a computer monitor and look up families by name. They can then choose pictures of members of that family. This is just one example of the use of video discs and computers in communication.

As you have read this chapter you will note that communication can be described as information processing. This is probably a new term to you. It means that information is processed or that something is done to it to make it more usable. If you have something to say to hundreds of people, it may not be easy to tell it to each person individually. If you can put this information into printed form, each person can read it. You have processed the information. You have made it more usable.

Computers are information processors. They are used to organize and manipulate information. In this way they are communication devices. We usually think of computers as only working with numbers. Though this was true at one time, it is no

To answer a customer request, this utilities company clerk can enter the customer's code into a computer and, instantly, billing records and other information will flash onto the terminal screen.

longer true. Modern computers work with numbers, words, and pictures. See chapter 9 for detailed information on computers.

There are two classes of computers: analog and digital. Analog computers measure rather than count. They are used to simulate actions. Analog computers are used in machines such as flight simulators to train pilots and astronauts. A slide rule is another example of an analog computer. Numerical values are represented by distances on the rule. By adding or subtracting distances, simulated numbers can be added or subtracted. Analog computers make up only about 10 percent of the computers now in use.

Digital computers are used to count. People put information into the computer. They then tell the computer how to process this information. This is called a program. The program is a set of instructions that tell the computer what to do with information.

Working in Communication

Education Requirements
Education and training needed to find employment:
1. High school education
2. Technical school training in areas of work: pressman, draftsman, or electrical technician.
3. Community college, two-year associate degree in areas such as broadcasting engineer, journalism, or commercial art.

left The person who puts together the film for printing is called a stripper. *right* A cameraman shoots the nightly news.

Communication

Suggestions for Further Study

Experience

As with many jobs, experience in the field is required. So how does one gain experience if no one will hire you? Gaining experience can come in many ways, for example, a tough job like that of a television camera operator. A simple way to get experience is to volunteer your services to your high school or college media department in order to learn the fundamentals. Next, try the local educational television station for possible volunteer work while you learn your trade.

Volunteer work will always help, especially when one needs a good recommendation for that important first job in your field. Remember to ask other workers how they got their start, how they like the work, and the conditions under which they must work. These answers will help you decide your future.

Observation

Look around you. What examples of communication do you see? Make a list of all the examples you see. Be careful! You will find examples of communication in the strangest places. Have you checked the kitchen in your home? How about a nearby street corner?

Some methods of communicating have changed little for hundreds of years. Can you find any of these? What is the newest method of communicating you can find? What methods do you know that are no longer used?

Involvement

You can get involved in communication in many ways. The hobbies already mentioned are a good way. Many schools have camera or photography clubs. There may be one in your community.

Does your school have camera or television equipment? Perhaps you can help operate it. Does your school have a stage crew? If so, they are working the sound and light systems. They may be looking for volunteers.

What groups in your community are working with communication? You will find that almost everyone is. Do you belong to a youth or scout group that needs help? Are there posters that need printing or papers that need to be delivered?

Did you ever think that the person who delivers the newspaper is working in communication? Certainly printed communication must be distributed to the readers.

Questions

1. What are some forms of personal communication?
2. What methods of communication did the American Indian use? Which was the fastest?

3. Which medium, radio, television, or newspaper has the greatest impact on us? What reasons can you give?
4. Name some ways human beings communicate with machines.
5. How do machines communicate with people? Machines to machines?

Key Words

communication
source of information
channel or carrier
receiver
noise
feedback
photography
camera
f-stop adjustment
offset printing

letterpress printing
telecommunication
microwave
broadcasting
producer
script
director
lens
shutter
lithography
interactive video

Producing a Radio Script

Concept: *Mass Communication*
The ability to communicate is very important to technological progress. People can communicate with other people. People can communicate with machines. Machines can communicate with other machines. Communication on a large scale is called mass communication.

The radio is a popular means of communication. Even with the development of television and other electronic communication media, millions of people still listen to the radio.

The radio is an individual medium of communication. It is geared toward one person. The writer for radio has to get the listeners' attention with humor or interesting dialogue. There are no pictures to attract the attention of the listener, who is often doing something else while listening.

Objective
In this activity you will write a short radio message. You will write a script and act it out with some of your friends. You can record the message and play it back as if it were being broadcast on the radio. You may be able to broadcast your message over the school public address system during school announcements.

Procedure
All you will need for this activity is a tape recorder with a microphone. A simple cassette recorder will work fine. If you or your school have larger, more sophisticated equipment, you could use it. The larger recorders may allow you to edit and dub in sound tracks from other recordings.

To begin writing your radio script:
1. Select the type of program you will produce: music, drama, news, commercial.
2. You may want to start with a 30-second commercial or a short public service announcement.
3. Write the script. Some hints for writing the script are given below.
4. Practice the script. If sound effects are to be used, you may have to build the apparatus for them. All actors in the program should practice their parts until the program sounds as good as possible.
5. Record the program.

The style of radio messages must create vivid images in the mind of the listeners. A personality must be created to attract and keep the listener's attention. Repetition is often used for emphasis. The writers for radio must remember that they are writing a message that will be heard but not seen. Words must be concrete, vivid and specific. Abstract words and phrases must not be used. Be aware of the sounds of words.

Practice the message using a tape recorder. Record the message then play it back. Listen to the sounds as well as the words. Varying sentence length will help reduce monotony. Changing the volume of the sound can spotlight a particular message. Change in volume can be used for emphasis.

Time the message as you read it. In radio, all messages must be timed very accurately. If you write a commercial for a 30-second time spot, it must be exactly 30 seconds long. You can estimate about 70–80 words in a 30-second commercial. A 60-second commercial will contain about 135–145 words.

The basic materials of a radio message are sound effects (abbreviated in the script as SFX), music and voice. Include all three in your message.

Sound effects are easy to create. Rattle a piece of sheet metal to sound like thunder. Water poured into a pan from a can with holes in the bottom will sound like rain. Footsteps can be created by stamping your feet on the floor. Shake sleigh bells to create a winter or snowy setting.

Music can be used in the background. Sometimes the music can set the theme or atmosphere. Loud, fast music or soft and slow music set different moods.

Be careful that the voice is not monotonous. Volume should change. There should be changes in emphasis on particular words. The script should not sound as if it is being read by the actors. Practice will help eliminate the sound of a script being read. It should sound natural.

Here is a script that was written as a commercial for popcorn. It is a good example of the form and style of a 30-second commercial. It requires an announcer, three actors and a background crowd. Sound effects could be produced by persons in the crowd.

Client: POPSIT Corn Company
Length: 30 Seconds
Date:

(SFX: PERSON EATING POPCORN LOUDLY)

BILL: Hey, buddy, keep it down. The movie's about to start.

JIM: Sorry, guy, but I can't help it. Popcorn just makes a lot of noise. But what do you expect? Fun things always make a lot of noise. Have you ever heard of a party that wasn't noisy? Or a circus? Popcorn is too much fun to eat; you can't keep it quiet. (yelling) Hey, everybody. Popcorn is loud, isn't it?

CROWD: (Yelling) Yeah!

JIM: (Yelling) But it's so much fun to eat, we don't care, do we?

CROWD: (Yelling) No!

JIM: (Yelling) Everybody—crunch!

(SFX: CROWD EATING POPCORN LOUDLY)

KAREN: Where are you going, Bill?
BILL: If you can't beat 'em, join 'em.
ANNCR: Join the fun. Eat popcorn.

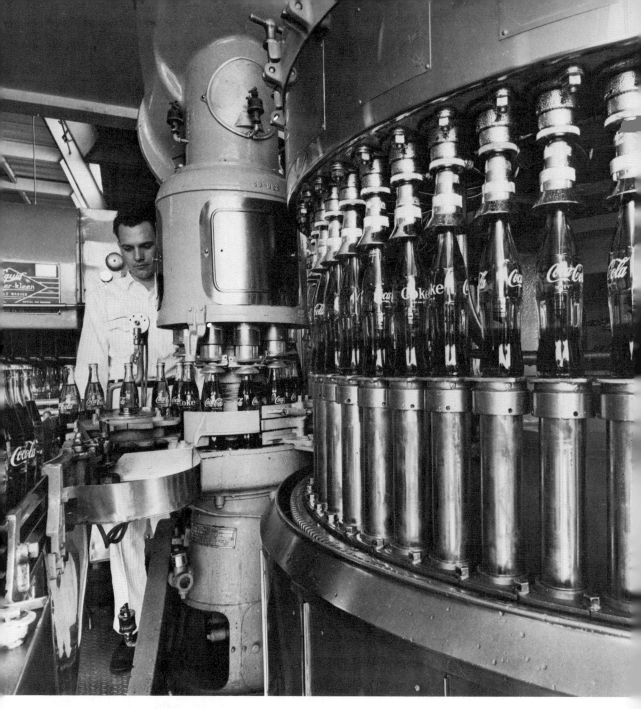

The bottling of soft drinks is an automated operation. The bottles are filled and capped by machine.

Chapter 8 Automation

Introduction

What is automation? One easy way to describe automation is to say that it is automatic control. Total automation occurs when a process is completed without human control. A process rarely occurs without any human control at all. Usually some human control is involved. But, for this study we will say that automation is automatic control. At the same time it will be understood that this means human beings may have some control of the process.

When a factory, assembly line, or some process is automated, machines are used in place of human labor. Parts are moved on assembly lines. A workpiece—work being manufactured—is fed into a machine automatically. Work is done on the piece, and it is moved to another machine. All this is done without human intervention.

Another way of describing automation is by saying it is automatic regulation. Again, an example helps. The thermostat in your home is an automatic regulator. It regulates the amount of heat in your home. As the house cools, the thermostat starts the furnace. The house gets warmer. When the house reaches the proper temperature, the thermostat sends a signal to the controls on the furnace and it stops.

Components of Automation

As has been done with other concepts in this book, automation can be broken into several smaller parts to help make it easier to understand. The five components of automation are: action, control, sensing, decision and program.

Motors, conveyors, and valves are all action components in an automated system. Energy is applied through them to move materials.

HISTORICAL DEVELOPMENT OF AUTOMATION

Time	Development
1642	Blaise Pascal's arithmetical engine
1788	James Watt's centrifugal or fly-ball governor. Regulates the speed and power of steam engines
1790	Oliver Evans' "automated" flour making factory. Mechanical devices move the raw materials through the various flour-making steps to filling the bags
1801	Jacquard's automated loom. Movements of many shuttles are controlled by punched cards to produce intricate patterns in the cloth

continued

Action Components

Action components are those parts of an automated system that apply energy to move materials in the system. Energy can be applied in several forms, such as heat to change the temperature of a material or in the form of electricity to run motors. The motors may drive conveyors, which move materials. The motors also may do work such as drilling holes or cutting material to size.

Control Components

The action is carried out because controls regulate them. A switch is closed, electricity flows, and the motor runs. The switch is a control. At the proper time the switch opens and the motor stops running, or a control causes a valve to open allowing a liquid to flow into a tank. At the proper time, the control can be activated to cause the valve to close, stopping the liquid from flowing.

Sensing

Sensors are placed in the automated system to detect and measure how the system is operating. These sensors are connected to indicators such as dials or gauges. A *thermocouple* is placed in a pipe to detect and measure the temperature of the liquid flowing through the pipe. The temperature is indicated on a gauge we call a *thermometer.* The speed of a motor is detected

HISTORICAL DEVELOPMENT OF AUTOMATION

Time	Development
1822	Charles Babbage's analytical engine. Eventually leads to the development of computers
1884	Hollerith's census machine
1923	Capek wrote the play R.U.R.
1946	D.S. Harder of the Ford Motor Co. originated the word "automation"
1946	First generation of computers begins
1950	Norbert Wiener publishes *The Human Use of Humans* —predicts automatic controls would make the automatic factory a reality within 25 years
1956	Work by Engelberger and Devol leads to the formation of the Unimation Company
1960	Three chemical plants under digital computer control have been built
1960	First industrial robot built
1961	Unimation company builds "Unimate" its first industrial robot. Used in auto assembly plant
1969	Computers begin to be used to control industrial processes in Britain
197?	PUMA—first robot designed exclusively to assemble the parts of computers and motors
1986	An estimated 26,400 robots have been installed in the United States. About 1,685 of these have vision capability.

and indicated on a dial showing revolutions per minute (RPM) of the motor. This is called a *tachometer*.

With these three components, (action, control, sensing) automation has not actually taken place. There has only been *mechanization*. The machines have been used to do much of the work for humans. But, control of these machines is still in human hands. Energy is being applied and materials moved or put into position. Controls regulate the machines. Sensors measure how well the system is operating. But, humans must make the decisions to activate the controls.

In the mechanized system, human beings read the information shown on the dials and gauges. Based on this information, they decide what adjustments are needed and then make them.

For example, sensors indicate that a motor is running at a certain speed. The operator sees this and compares the reading with a figure known to be the proper speed for that motor. Based on this information, the operator decides to change the speed of the motor. A control is activated and the speed of the motor is changed.

In this example the information about the proper speed of the motor was stored in the operator's brain or in an operation manual for the machine. The operator made the decision and then made any adjustments needed. In a completely automated system, a machine (a computer) replaces the person in the decision-making process. Information is sent from the sensors to the computer. Information about how the system should be operating is stored in the computer's memory. Information from the sensors is compared with the information in the memory. Based on this comparison, the computer makes a decision and instructions are sent to the controls. If needed, the controls are activated and adjustments made.

Decision and Program

Two new components have been added to make the automated system. A *decision* component receives incoming information, compares it with stored information and, if needed, sends commands to controls in the system.

The stored information or *program* is the fifth component of the automated system. The program includes both process and command information. Process information contains data about how the system should be operating. The command program includes instructions that direct the controls to do certain operations.

In the mechanized systems, the sensors indicated how the system was operating. People read these indicators, compared the information with data on how the system was supposed to operate and then made any adjustments needed.

In the automated system the person is replaced by some type of computer. The computer receives the information, makes the comparison, makes the decision, and commands the controls to activate.

This sending of information from the sensors, through the computer, and to the machine again is called *feedback*. Feedback is needed before automation is possible. Through feedback of information, machines become self-correcting and self-adjusting. No human control is needed.

Several examples will help explain automated systems. The heating system in your home is a good example of a simple automated system. In early, nonautomated heating systems people had to regulate the amount of heat put out by the furnace. When the house became cool, someone added fuel to the fire. More heat was produced and the house became warmer. The person adding the fuel had to be careful not to add too much or the furnace would overheat and the house would get too warm.

With an automated system, a thermostat senses the temperature in the house. It also contains a program and makes the decision to regulate the furnace. When the temperature in the house goes below the figure set on the thermostat (this is the program though it is only one bit of information), this is sensed. A signal is sent to the control on the furnace. The control is an electrical switch. The switch is closed and a pump forces fuel into the combustion chamber. At the same time a spark is produced, which ignites the fuel.

The furnace will continue to burn until the thermostat senses that the house is warm enough. Then a signal is sent to the control and the flow of fuel is stopped, halting the burning. This on-off action of the thermostat and furnace continues regularly to keep the temperature in the house within a comfortable range.

Systems of traffic lights used to regulate the flow of traffic in cities are another example of automation. The simpler systems are not automated. They are timed to change at set periods of time, for example, every twenty seconds. Each change is regulated by a clock timer that is preset. This is not an automated system because there is no feedback in it.

More complex traffic light systems are automated. Feedback is built into the system. At intersections where several streets come

An engineer adjusts the electronic eyes of a robot. The "eyes" can sense the position of parts, telling the robot where to move to pick up parts.

together, it may be necessary to allow traffic to flow for a longer period of time when traffic is heavier on one street than others. Sensors can be placed in the roadway. An automobile being driven on the little-used street runs over the sensor, causing the light for that street to change to green. After the auto passes, the light changes again, allowing traffic on the heavily-used street to move once more. In this way traffic can move in the heavily-used street until an automobile appears on the little-used street.

Another example involving you may help explain automation. Suppose you are riding your bicycle down the street. A dog starts to run across the street in front of you. You see the dog (your eyes are the sensors in this system) and this information is flashed to your brain. Here the information is received and compared with information stored there. Your memory (the program) says that hitting the dog can cause you to wreck and cause possible injury. So, the decision is made to apply the brakes on the bicycle and slow down to avoid the dog. This decision is flashed to your hands where the controls on the bicycle brakes are squeezed. The brake pads are pushed against the wheels and the bicycle slows.

You and your bicycle slow enough to avoid the dog. This information is fed back to the computer (your brain) where the decision is made to release the brakes and to continue down the street.

In this example there were all the necessary components of an automated system. Your eyes were the sensors. Your brain was the computer that received the information, made a decision based upon stored information, and commanded the controls (your hands and the levers on the bicycle) to activate the action elements (the brake pads).

Controls in Automation

To help understand automation, let's learn about a simple control. This is an example that is familiar.

First, look at the thermostat in your home. It is used to regulate the temperature in the house. How does it work? There are several types of home thermostats. One type uses a bimetallic strip to sense the changes in room temperature. The strip is made of two different metals which are fused. When the strip is heated, it expands. But, the two different metals expand at different rates. This causes the strip to bend.

When the room cools, the bimetallic strip bends so that two electrical contacts touch. This allows electricity to flow to the

Bimetallic Thermostat

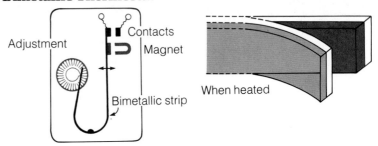

furnace control box. Fuel is pumped into the furnace where it is ignited and burned. Heat is generated and flows to the rooms in the house.

As the room gets warm, the bimetallic strip in the thermostat also warms. As it warms, it bends in the opposite direction until the contacts are pulled apart. The flow of electricity stops and the furnace is stopped.

Each time the house cools to a certain temperature, the thermostat signals the furnace to start. When the house warms to a certain temperature, the furnace is stopped.

Automation in Practice

As our study of technology continues, you will realize that this has been a very simple explanation of automation. But it should be enough to give a basic understanding. Automation is important to modern technology. It undoubtedly will be important in our future.

Automation can be found in nearly all areas of technology. Our first thoughts about automation may be of automated assembly lines with robots doing the work of humans. But, actually, automation can be found more often in other areas. The modern telephone system would not be possible without automated switching systems. Men were able to travel to the moon and return because the complicated launching procedures for the rockets and controls for the space vehicles were highly automated.

The following descriptions of automation in practice should help show its importance in all of technology.

Manufacturing

Large assembly plants such as automobile plants must have thousands of parts in storage. These must be readily available when needed. It is difficult to keep track of the parts and to keep

Welding, often a dangerous task, can be done automatically by robots.

an inventory of stock on hand. Automated parts inventory and handling systems are used to do this enormous task.

Using an automated plant warehousing system, a single operator can store and retrieve the parts needed to feed an assembly line. The system can store, retrieve, and maintain a running inventory on as many as 50,000 parts. The machine has over 5,000 cells or bins for parts storage.

To place parts into storage, the operator puts them in special containers or, if properly packaged, they are left in the shipping cartons. These are placed on a platform which moves throughout the many rows of bins. The bins run from floor to ceiling in the storage area. The operator enters information concerning the parts into the control console of the system. The parts then are carried to the storage area and placed into an empty bin.

When the parts are needed, the operator enters the code number for the needed parts into the console. The platform moves into the bin area, retrieves the proper box of parts, and delivers it to the operator. No human hands touch the box of parts between the time it is placed on the platform for storage until it is delivered to the platform by the machine for use in assembly.

In addition to storing and handling the parts, the system also maintains records of all parts that are stored. To make an inventory of parts on hand, the operator merely enters the proper

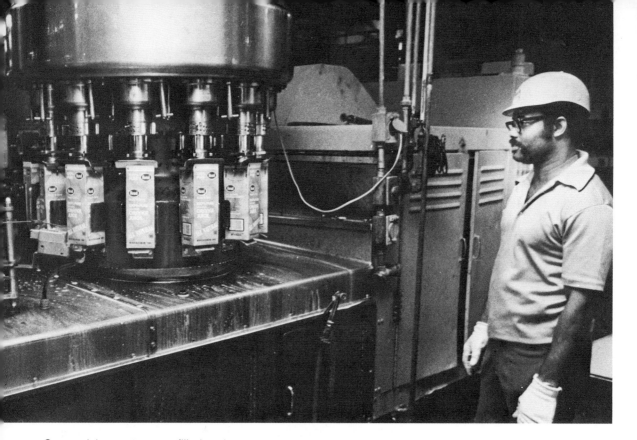

Orange juice cartons are filled and sealed automatically.

code in the console, and a typed list of all stored parts will be produced.

The bottling of soft drinks is another operation that is automated in several ways. The mixture of syrup and water is controlled automatically. Sensors are placed in pipes carrying the mixture. It measures the amount of sugar in the mixture very accurately. If the sugar content is low, a signal is sent to controls that open valves, letting more sugar enter the mixture.

Before filling, the drink bottles are sorted and washed. The bottles then move along the conveyor line and are stopped under the filling spouts. The liquid flows into the bottles with sensors measuring the correct amount and stopping the flow when the proper amount has been put into the bottles. The bottles are again moved along the conveyor where caps are added and cartons are filled. Often the only person working on this machine is an inspector who checks the filled bottles.

The machining of metal was among the first processes to be mechanized and then automated. Machine tools were first controlled by punched tape. There was no feedback. The tape merely controlled the location and cutting action of the cutter.

These machines are called numerically-controlled (NC) machines. Each hole (or combination of holes) in the tape represents a command to the machine. Sensors detect the holes. A message is sent to the controls on the machine causing it to move in a certain way. Magnetic tape similar to that used in tape recorders replaced most punched-tape in NC machines. Today these automated machines are controlled by computers. This is called Computer Numerical Control (CNC). CNC machines can store and transmit large quantities of very complex information. They are capable of doing very complicated tasks.

Feedback is present in modern automated machine tools. Any bending of the metal is detected by sensors. Vibration of the cutter or work piece can be detected. This information is fed into a computer, which makes any adjustments needed to produce the best possible cut. Even very small changes needed because of tool wear can be made. The result is a higher quality finished product.

In some areas of manufacturing, the flexibility of the human hand is still needed. This is true in assembly industries such as automobile production. It is difficult to make machines that can handle a variety of parts on an assembly line. Though highly mechanized, these machines usually lack the feedback component necessary for true automation.

Chemical processing is among the most highly automated industries. One automated refinery, for example, can produce many different materials, all automatically. Computers provide analysis and control functions. Thousands of sensors check factors such as temperature, pressure, flow rate, and liquid levels. These can be monitored and controlled from one control room.

In such a highly automated system, if a factor exceeds set limits, a computer sounds an alarm and prints out a message telling what the problem is and where it is located. If needed, the operator or a technician can make the necessary corrections.

In paper making, automated systems are necessary to produce high quality paper at minimum cost. Computers check over 200 factors in the process. Each factor can be controlled, if needed.

Look around for examples of automation. What can you find? Street lights that come on when it gets dark and turn off at daylight? A burner on the kitchen stove that holds a set temperature? An automatic door opener at the grocery store? A tape player that turns off when the tape reaches the end? These are only a few examples of simple forms of automation. There are hundreds that can be found. Do you know which are examples of true automation and which are only mechanization?

Robotics

Robots are an important feature of automation. A robot is a machine that can be reprogrammed to do different jobs. Many people still think of robots as machines that look like humans and do things the way humans do them. This is not a true image of robots. Actually robots take many forms depending on the jobs they do.

The first idea of a robot was to replace humans as workers. In 1923, Karel Capek, a playwright from Czechoslovakia, wrote a play titled "R.U.R." (Rossum's Universal Robots). In the play a scientist named Rossum invents a substitute for flesh and bones. With this he hopes to create artificial life forms. He fails but his son succeeds in making the perfect manual labor machine. His robots are very successful and appear to be taking over the world. Eventually the robots fail and humans gain back control. The word "robot" came from the Czech word "robota" used in this play.

Modern robots were developed by Joseph Engelberger and George Devol. Their work led, in 1956, to the formation of the Unimation Company, one of the first manufactuers of industrial robots. These robots did not resemble humans. Only the human arm was copied.

Think of a robot arm in human terms—arm, hand, and wrist. These terms are used to describe parts of a robot because they work similarly to the human arm, hand, and wrist.

This industrial robot is made with a long arm to reach into the truck bed. The robot can swivel around, the arm can move up and down, the wrist can move in and out and turn around. This robot has how many degrees of freedom?

The rear cage of this robot swings open to allow access to the computer. Notice the disk drive in the robot's side to allow input of different programs.

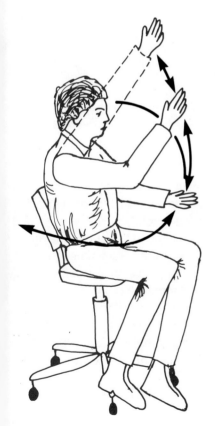

To understand the movement of a robot, pretend that you are a robot. Sit on a swivel chair. You and the chair are like a simple robot. Swivel (turn) around in the chair. In robotic terms, you have one degree of freedom—you can move in one direction. You can move around an axis which is the swivel point on the chair.

Extend your right arm. This is like the arm on the robot. Move your arm up and down without bending your elbow. You now have a second degree of freedom. In addition to being able to turn around, you can now reach out in front of you and lift objects. These two degrees of freedom make it possible for you to do more work than you could do with only one degree of freedom.

On the end of your arm is a hand. It can open and close to grip objects. You now have a third degree of freedom. As a robot, you are able to do many things. You can grip an object, lift it up and place it to the left or right. Add an elbow and you have a fourth degree of freedom which makes you, as a robot, much more flexible. You can now move the object up and down, left and right and in and out. With these four degrees of freedom, you are a very good robot, capable of doing many tasks.

Robot arms

The movement of a robot arm is usually in a circular path. Just as your waist or the swivel in the chair rotates, a robot can swivel around vertical axis. This is called "armsweep." The robot arm also

Automation

Robot Work Envelopes

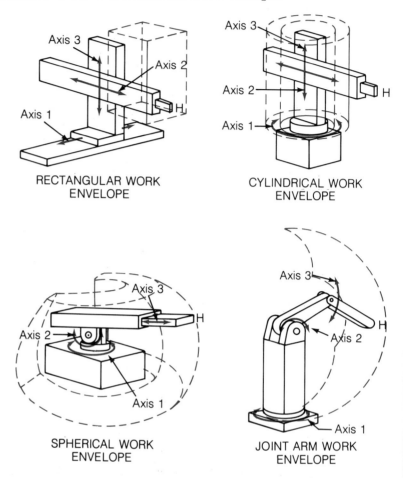

RECTANGULAR WORK ENVELOPE

CYLINDRICAL WORK ENVELOPE

SPHERICAL WORK ENVELOPE

JOINT ARM WORK ENVELOPE

can move vertically — up and down (shoulder swivel). If a joint similar to the human elbow is built into the arm, the robot's ability to reach is increased. This is called "pitch."

Many different devices can be attached to the end of the robot arm. This is usually called the "end effector." Sometimes the end effector is like a human hand. It can grip objects. It can move left or right (yaw). It can move in a circular motion (roll). Grippers can be designed to grip many different shapes. Often the end effector can be designed to do only one job. The gripper is designed so that it can open and close around an object. The gripper has enough grip or closing pressure to hold the object tight enough to lift it without dropping it.

Watch your hand as you pick up a pencil to write. Notice the movement of each finger. Though robot "hands" are not as complex as your hand, they can be made to do special tasks such as grip a pencil or a tool. They can be made with a spray gun on the end of the arm instead of a hand. A welding electrode holder can be installed or a wrench can be used to tighten a bolt. Often the device on the end of the arm is a special tool rather than something that looks like a human hand.

The robot has a computer for a brain. When you were sitting in the chair pretending to be a robot, your brain was controlling your motions. In the robot, the computer is the "brain" doing the controlling. Humans have programmed the computer to make the robot move. The robot cannot think through the job and tell itself what to do. It can do only what a human has programmed into the computer.

Types of Robots

There are many ways of classifying or grouping robots. Each way tells something about robots and the way they work. One way is to group robots into the kinds of work they can do: assembly, material handling, welding, and spray painting. Often robots are classified by the type of motion they go through: "pick-and-place," "point-to-point," and "continuous path."

Robots are also described as servo and non-servo controlled. The non-servo controlled robots are fitted with mechanical stops to limit their travel. For example, there may need to be a limit to the distance the robot can swivel around the vertical axis. This could prevent cables from wrapping around the machine, for example. In a non-servo controlled robot, the stops are a block that the arm bumps against to stop it at the proper point.

To help visualize this, think of a model electric train running on a straight track. The train needs to be stopped before it runs off the end of the track. A brick could be place near the end of the track to stop the train. The brick could be moved to different positions on the track to stop the train at different places. This would be like a non-servo control.

Non-servo controlled robots are generally less expensive. Many can be made using simple and inexpensive controls. Usually non-servo controlled robots are smaller. They can be made to move and stop very accurately.

If, instead of the brick, the train was controlled by slowing it down with the speed control, then turning off the electric current when the train was at the spot where it was supposed to stop, this would be similar to servo-control used in robots.

Robots are used to paint auto bodies. The automatic painter eliminates the need for workers to perform difficult jobs in a dangerous environment.

Servo-controlled robots are made in many shapes and sizes to do a wide variety of jobs. In general, servo-controlled robots can be used to do many more jobs than the non-servo controlled robots. Control of the servo controlled robots is done by the use of computers and microelectronics.

Working Robots

On the job, robots are also described by the type of power used in them and their "work envelope." Generally, three types of power are used in robots: electrical, hydraulic, and pneumatic. Pneumatic power is most common in robots used for picking and moving small, lightweight parts. Pneumatic cylinders are used to provide the power needed to move the different parts of the robotic arm.

Hydraulic cyclinders are used in larger robots to provide the power to lift and move larger, heavier objects. Hydraulic and pneumatic powered robots are used in jobs such as spray painting where sparks from electric motors could cause explosions.

Many small, hobby robots are electromechanical systems. This means the electric motors are combined with mechanical systems of levers, pulleys and belts, chairs and sprockets, etc. Often there is a small electric motor to provide power and to control each movement of the robot. Each motor is controlled to control the overall movements of the robot.

Often electric motors are combined with hydraulic and pneumatic systems in the largest robots. This combination results in a robot that can be controlled better or is more suited for a special use.

Work envelope refers to the area within which the robot can reach. When you sat in the chair pretending to be a robot, you could reach only so high, out so far and could swivel around only so far. As a robot, you could work in only a limited space. This was your work envelope.

The work envelope for a robot depends on its design. Some robots are designed so that they have a rectangular work envelope. These usually have an in-out, up-down, right-left motion. They have three degrees of freedom.

Robots with cylindrical work envelopes may also have three degrees of freedom. They can move up and down, in and out, but turn around like the swivel chair.

A robot with a spherical work envelope swivels around, and the arm pivots up and down and moves in and out.

Jointed arm robots have the most complex working envelope. The jointed arm looks very much like a human arm. It is made so it can rotate around a base. It can have a shoulder rotation, an elbow and wrist. The many joints can provide many degrees of freedom to the robot. The jointed arm design provides the largest working envelope of any of the robot designs. This design requires more complex control because of the many joints and possible areas of reach.

Effects of Automation

When a factory or assembly line is automated, robots are being used. The robots are used in place of human labor. Parts are handled automatically. They are moved on conveyors rather than carried by humans. Machine operations are controlled by the machine. The piece of work is fed into the machine, work is performed on it, and it is passed down the line where more work is performed. No human labor is needed.

One can see from this example that fewer people are needed to work in an automated factory than nonautomated ones. Automation does away with work that is boring to human beings. When a job must be done over and over again, it becomes boring. People like variety in their work. Drilling a hole in a piece of metal is a good example. A worker has to put the metal into the machine. A hole is drilled. The piece is taken out of the machine. Another piece is added. A hole is drilled. This simple operation is repeated hundreds of times in a working day. This may become very boring for the worker.

This operation can be automated. The machine can be designed so it can do all the operations. No human beings are needed to run the machine. A worker does not have to watch the machine. It is automated so that it controls itself. Does this mean that the worker has been put out of a job by automation? Not necessarily. The worker can do work now that requires human control. The worker now may look after several automated machines. The worker can spend time on creative work that requires the use of human mind and hands.

Even automated machines need to be adjusted from time to time. The machines need to be adjusted for different operations. There is a limit to the number of produced parts needed. When a production run is finished, the machine must be reset to do other operations.

Some workers still may lose their jobs because of automation. But, automation often creates more jobs than are eliminated. People are needed to make the machines and to repair them. Computer operators are needed. All these are jobs that did not exist before automation.

These are also jobs that need highly skilled people. The workers who lost their jobs may not have these new skills. In some

Automobile assembly lines are among the most automated in manufacturing. Robots are rapidly being used to do work formerly done by people.

cases they can be retrained. Others have to move on to other jobs. So, automation creates a need for highly skilled people. Often more highly skilled people are needed than those who have been replaced. In such cases automation actually may increase the need for workers. Jobs have been created as well as eliminated.

Automation can speed production. We have already seen how production can be increased. One worker can produce as much as several workers using nonautomated machines.

Automation can make the work easier for the workers. The boring and tedious work can be done by robots. Dangerous work can be done by machines. Very accurate work can be done by machines. The automobile engine is an example. In the engine plant, one machine does 105 operations in 38 seconds. Another machine works on 17 engine blocks at the same time. Each operation is done with high precision. These machines are automated.

Automation saves time. Work can be done faster by automated machines. A modern automobile can be assembled in about thirteen hours. This is from the time the first part starts down the assembly line until the auto is driven from the plant.

There are also problems with automation. It already has been seen that automation seems to eliminate less skilled jobs. Often workers are replaced by automated machines. Many of these workers cannot be trained in the higher skills needed.

Though automated machines and robots are expensive to build, their use can result in big savings for a company. One Japanese company substituted a computer, a robot team and four humans for 120 human workers. They were making vacuum cleaners. The company reported a 2900% improvement in productivity. Fewer workers were used and the company reduced the cost of making the cleaners.

Reduced cost is one of the great benefits of automated production. Robots can work day and night. They do not call in sick. They are always on the job. It is estimated that an industrial robot will pay for itself in about 3-5 years.

Robots are becoming more sophisticated. Some can work on a constantly moving assembly line, doing several tasks at once. When something goes wrong some robots can search their own systems to find the problem. Some can even correct simple problems and then go back to work.

It must be remembered that robots are machines. They are made by humans. They can break down. Then need repairing and adjusting. They often need to be reprogrammed or redesigned to do new tasks. They wear out and eventually must be replaced. All this requires humans to do the work. Humans have not been totally replaced by robots. In some places robots can be found to be doing all the work with no humans in sight. But, humans have had a hand in this and will be needed.

Suggestions for Further Study

Observations

As you study this chapter on automation, look around you. Can you find examples of automation in your home or community? Where is robotics used in your community? Is there an industry that manufactures or uses robots? If so, you may be able to visit the industry. Can you obtain literature from a robotics industry. Look in magazines in a library for companies in the robotics field. Many of these will be happy to send you information.

You may be surprised that examples of automation can be found in your home. What about the automatic washer or dryer? Are they automated? Do they go through the wash or dry cycle automatically or do they have to be manually run through the cycles? If you do not have these appliances in your home, go to a laundromat. The machines there will be automated.

Is the dryer set to run for a certain length of time or does it stop when the clothes are dry? If it stops automatically, it probably has a sensor which detects moisture inside the machine. When the moisture level drops below a preset point, the dryer stops.

There are many more examples of automation in your home. How many can you find? Hint: look in the family or entertainment room, the kitchen or the garage.

Personal Involvement

You can become involved with automation. In chapter 4, Manufacturing, it was suggested that you manufacture a product in your school lab or at home. You can automate your production line. Build conveyor systems. Automate these conveyors using computers and robots. You can use a simple robot to pick and place objects on the conveyor. You can use a sensor to count products as they pass by on the conveyor.

Machines can be operated automatically. Jigs and fixtures can be built which hold parts in proper position on the machines. Computers and robots can be used in many ways in your production line.

Questions

1. Define automation.
2. Define a robot.
3. Why is automation necessary in modern industry?
4. How does automation affect humans?
5. How has automation affected you today?
6. Trace the history of automation from its beginning. When do you think automation began?

Key Words

automation
automatic control
bimetallic
conveyor
degree of freedom
detector
end effector
microelectronics
production
regulate
relay
robot
robotics
servo
solenoid
thermostat
work envelope

Robots

Concept: *Automatic Control*

Automated machines, including robots, are designed to operate without humans to control them. Punched tape, magnetic tape and computer memories contain the information needed to tell the machines what to do. Often the machines are capable of sensing conditions, receiving feedback and reacting to the conditions.

Robots are now an important part of automation. Many manufacturing industries are finding that they must use robots to stay competitive. Automated systems are being used to manufacture complete items without human hands touching them. Robots are replacing human workers in hazardous environments such as painting and welding.

Objective

You can learn about simple robots by building one or more of the robot kits now being sold. Some of these robots can be operated by remote control; others can follow a black line in the floor or be controlled by loud sounds. Infrared light beams can control some robots. Others can be programmed with a teach pendant or even with your microcomputer.

Several robots, available as kits, are pictured below. You can learn a lot about electronic circuitry while building these robots. You also can learn how simple robots work. Talk to your teacher about sources of robot kits. You may find some in local stores selling electronics supplies.

Sound Operated Robot This speedy little robot will bang into walls, but each time it does, it reverses itself and goes in a different direction. A small microphone on the front picks up loud sounds such as handclapping. The sounds cause the robot to turn and then move off in a different direction.

Programmable Robot You can program this robot to turn, pause, sound a buzzer, light a light emitting diode (LED) and run through the sequence several times. The brain of the robot is a small integrated circuit. You can program the robot using a teach pendant similar to the ones used on large industrial robots. Programming also can be done with a microcomputer through a special interface unit.

Both robots and the interface unit can be purchased in kit form. They are compact with lots of small parts to be assembled. The electronic circuitry is soldered onto printed circuit boards. These kits require some skills with tools. The kits make interesting class projects. They also are excellent projects for electronics and robotics hobbyists.

Automation

Computers are essential in transportation to making everyday decisions. This computer is being used on a locomotive.

Chapter 9   Computers in Technology

Introduction

Do you have a computer in your home? How many computers are there in your school? How long have computers been used in your school? Quite likely computers have been used in your school for only a short time.

You have probably seen many changes that computers have made in your school and your home. The computer has had greater impact on humans than any other modern machine. Since electronic computers were made in 1964 using integrated circuits, there have been dramatic changes in our world.

Today computers are used in home appliances. Computers are used to control the running of automobile engines. This book was written on a computer and the type was set on a computer. Just a few years ago computers were large machines used by only a few researchers. Today they are used by many people even for playing games.

Development of Computers

The technology that has made microcomputers possible did not happen suddenly. It was developed over a long period of time. Machines that help people calculate can be traced back as far as 3000 BC when the abacus was in use. The abacus was a frame with wires that held rows of small beads. These beads could be moved around on the wires. By moving the beads they could be

Modern computer systems are quite different from early calculators.

used as counters to keep track of numbers when adding, subtracting, multiplying, and dividing.

The abacus is still used in some countries. Once a contest was held to see who could calculate faster, a man using an abacus or one using a pocket calculator. The man using the abacus won the contest.

The slide rule is another kind of calculating machine. About 1620 Edmund Gunter invented a simple rule to be used in calculating. Distance on the rule represented numbers. By measuring these distances with dividers, math calculations could be made. In 1850 the Mannheim rule was developed. Modern slide rules are similar to the Mannheim rule. Instead of measuring with dividers, a slide could be moved to make math calculations.

Until recently the slide rule was a basic tool of the engineer. Today it has been replaced by the electronic calculator. The modern calculator is quicker, more accurate, and similar in cost.

Math tables were used to simplify complicated hand calculations. Like the abacus, these tables of numbers have ancient origins. Astronomers used math tables to calculate the movement of heavenly bodies. Mathematicians used them to calculate square roots and other math functions. It was not unusual for an engineer to own a small library of books of math tables. These books of tables required an unbelievable amount of time and work to develop. Today some business people still use tables to make calculations such as insurance rates and interest rates on loans.

Early calculators were developed for use in calculating math tables. A person could develop the math tables much faster using

The abacus is an ancient calculator which historians trace to Egypt, India and Mesopotamia.

298   Exploring Technology

HISTORICAL DEVELOPMENT OF COMPUTERS

Time	Development
3000 B.C.	Abacus in use
1614	Logarithms invented by Napier
1621	Slide-rule invented
1623	First calculating machine—Schickard
1642	Pascal's calculating machine
1820	Thomas Arithmometer—first mass-produced calculating machine
1822	Difference engine begun by Babbage
1884	Hollerith's Census Machine
1892	Burroughs adder-lister
1896	Holerith Tabulating Machine Company formed (eventually became IBM)
1944	First program controlled computer—Harvard Mark I (had 3/4 million parts)
1946	First generation of computers begins (used electronic tubes) ENIAC—First electronic computer (18,000 tubes, weighed 30 tons)
1948	First stored program computer—Electronic desk calculators replaced mechanical calculators
1951	First commercial data processing computers—LEO and UNIVAC
1956	FORTRAN—most influential computer language
1958	Second generation of computers begins (transistors used in place of tubes)
1964	Third generation of computers begins (integrated circuits used in place of transistors)
1965	First commercially sold minicomputer
1971	microcomputers
1972	Pocket calculators
1972	"Computer-on-a-chip"
1987	32-bit microprocessor chip—peak execution rate of 25 million instructions per second.

The Pascal calculator could only add numbers. Numbers were registered on the telephone-like dials.

a calculating machine. The tables also were more accurate. As calculators were improved they replaced the tables in math calculations.

An understanding of early calculators helps us understand the importance of modern computers. We can see how the modern computer developed. As early as 1642, an adding machine was invented. Blaise Pascal, a Frenchman, at the age of 19 made this early machine. It could only add numbers, but it was a great invention. A few years later Leibnitz, a German mathematician, made a machine that could multiply numbers.

In the late 1700's a Frenchman, Joseph Marie Jacquard, made a machine that used cards with holes in them to control looms. The punched cards were a kind of memory. The cards were used

Computers in Technology

The Jacquard loom was controlled by punched cards. The holes in the cards were a simple memory system.

Even after twenty years of work, Charles Babbage's difference engine was never completed. Babbage is recognized as the pioneer of the digital computer.

to control the loom to make intricate patterns in the woven fabric. The cards could be used over and over to duplicate the pattern.

In 1824 Charles Babbage proposed what he called a "difference engine.". This machine was made up of many wheels and gears. When cranked it could perform complicated math calculations. Though it did not work, it was a good idea. Later other people were able to use the idea and improve on it. In 1852 George Scheutz made a difference engine that worked. It was used in making math tables.

At about this time an American, Herman Hollerith, used Jacquard's idea of the punched cards to make a machine for calculating the 1890 U.S. Census. It had taken more than seven years to report all the figures from the 1880 census. Using his new machine, Hollerith was able to report the 1890 census in just over two years. Hollerith's Tabulating Machine Company was the first computer company and eventually became the International Business Machine Company (IBM).

These early machines did not record the calculations. In 1888 William S. Burroughs patented a printing adding machine. Figures could now be printed out on paper and kept for record.

Genealogy of Switches in Computers

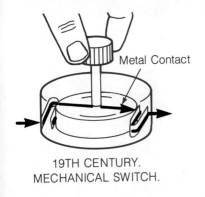

19TH CENTURY. MECHANICAL SWITCH.

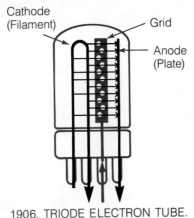

1906. TRIODE ELECTRON TUBE.

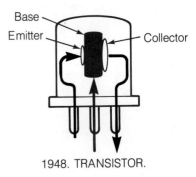

1948. TRANSISTOR.

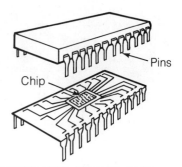

INTEGRATED CIRCUIT "CHIP"

These machines were still not what we now call computers. Adding machine is a better name for most of them. Calculator became the name when the machines were made to add, subtract, multiply, and divide. They were mechanical devices. They were made of many gears, cranks, and levers and were hand powered. People had to set figures into their number wheels. A crank was turned to turn the wheels, and the results were read off the numbered wheels.

Early Computers
The earliest computers were quite different from the ones we know today. They were very large and contained many thousands of parts. They were not electronic computers but used telephone relays and other mechanical switches.

One of these early computers was the Harvard Mark I, first used in 1944. This machine was 51 ft. (15.5 m) long and 8 ft. (2.5 m) high. It contained more than 750,000 parts. It was able to store seventy-five numbers and could add them at the rate of three per second.

The Mark I has been called the first automatic computer. It was driven by a long rotating shaft powered by electric motors. It re-

The Mark I computer was very large and slow compared with modern computers.

Programming early computers was done by re-wiring circuits in the machines.

ceived instructions by punched tape based on an idea similar to the punched cards used in Jacquard's loom.

Many advances in computer technology were made during World War II. The need for fast calculations created a lot of interest in computer research. Mechanical calculators were used to break codes, calculate math tables and aim guns. They were used in engineering and scientific research projects. One of these was the Manhattan Project which developed the atomic bomb.

First Generation Computers

The Mark I and other electro-mechanical computers did not last long. Researchers were developing computers using electronics.

Electronic computers began the true computer revolution. The first generation of modern computers began after World War II was over. ENIAC (Electronic Numerical Integrator and Computer) was first used in 1946. Electronic tubes (similar to radio tubes) replaced the telephone relays. ENIAC contained 100,000 parts; 18,000 of these were tubes. It weighed 30 tons and took up 800 sq. ft. (74 sq. m) of floor space. ENIAC was more than 1,000 times faster than the Mark I.

ENIAC could store only twenty numbers. Programming involved rewiring the machine, which took more time than was needed to make the calculations. Later development of memories for computers solved these problems.

ENSAC was put into operation in 1949 in England. It could store up to 500 words. It could perform about 600 operations in one second. This made EDSAC more powerful than ENIAC. It has been estimated that in a few minutes EDSAC could do calculations that would take a human a year to do. In 1958 EDSAC 2 went into operation. It was fifty times more powerful than EDSAC.

A well known commercial computer of this time was the UNIVAC, first used in 1951. This was one of the first computers designed for commercial data processing. A new feature of the UNIVAC was the use of magnetic tape to store data. Soon magnetic tape had replaced the slower punched tape or cards for data storage. One roll of tape could hold as much data as thousands of punched cards.

Second Generation Computers

The second generation of computers began in 1958 when transistors replaced electronic tubes. The transistors were much smaller than tubes. Very powerful computers could be made much smaller.

Electronic vacuum tubes replaced electric relays in computers. This made stored programs possible.

Computers in Technology

Transistors and printed circuits permitted greater speed and better reliability in computers. This new technology replaced vacuum tubes for stored programs.

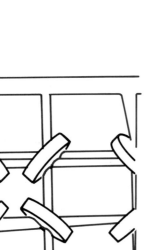

Part of a core memory, highly magnified.

Memory core magnets were strung on a grid-work of wire.

They also could be built containing up to one million transistors. The pea-sized transistors used less power and gave off less heat. The compact computer was now possible.

The core memory was another development during this time. Earlier memory systems were slow and unreliable. The core memory improved this.

Core memory was made by stringing small magnetic ceramic rings on a wire mesh. The rings could be magnetized in a clockwise or counter-clockwise direction. In this way a single bit of information could be stored on each ring.

Third Generation Computers

When integrated circuits or "chips" replaced transistors, the third generation of computers began. There is now a wide range of sizes

of computers from the small home microcomputer to the huge "super computer."

The new super computer can perform millions of operations per second. Thousands of terminals can be connected to these large computers. When used in this way, they are called "main frame" computers. Many people can do work on them at the same time. Each person needs only a terminal (keyboard and monitor) at each station. These are connected to the "main frame" through a network of cables.

Modern computer systems use microcircuits for memory storage. These are capable of multiplying two ten-digit numbers over one million times per second.

Small computers also are a part of this third generation. Today portable computers are common. A business executive can carry a portable computer in a brief case. While traveling letters can be written, sales reports prepared and advertising campaigns developed. The work is entered on the portable computer. This information can later be transmitted to the home office with a modem link through telephone lines.

When the executive returns to the office, the letters can be ready for a signature. The sales and advertising reports can be ready for presentation to the board of directors. The executive's time, even while traveling, is much more productive.

Some people wonder when the fourth generation of computers will begin. It may have started already. New advances in computer technology are occurring very rapidly. Consider the short time be-

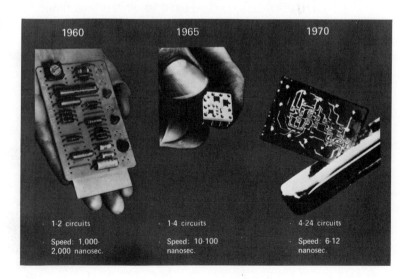

The progress of computers can be shown by the changes in size and speed of the electronic circuitry.

Computers in Technology

tween announcements of major advances or new computer products. Any one of these changes could be the beginning of a fourth generation. Most likely this new generation will not be the result of one new development. The new age will probably occur over a period of time and be more difficult to assign a definite starting date.

What Is a Computer?

A computer can be defined as a programmable machine that accepts, processes and displays data. To better understand what this means, let's look at a computer. We will use a typical home microcomputer as an example.

A typical home computer is a micro-computer system. System means that it is made up of several parts all acting together to perform a task. The computer is made up of three main parts. These are often called the hardware. These are (1) the computer itself — usually called the processor; (2) input devices such as the keyboard and disk drive; and (3) the output devices such as the monitor or printer.

The computer is an information processor. Information must be put into the computer in a form the computer can understand. It must be output in a form the human can understand. Information is stored in several ways. One of the most common ways of storing information is on magnetic disks. Other storage methods include magnetic tape and optical scan sheets. The punched cards used by Hollerith and Jacquard were older ways of storing information. The holes in these cards represented bits of information used to control machines.

A typical home computer is a system made up of the computer, monitor, keyboard and printer.

Exploring Technology

The system board or "mother board" is the central part of a modern computer.

The main part of a computer is the central processing unit (CPU). This is the brain of the system. Just like the human brain controls the body, the CPU controls the computer system. The CPU is a microprocessor that directs all the computer's activities. It is connected to a large circuit board called the system board or "mother board." All of the input and output devices are also connected to the system board. This connects them to the CPU.

A quartz crystal clock is an integral part of the CPU. When the computer is turned on, the clock begins vibrating, in some cases as fast as millions of times per second. The vibrations or pulses are very steady and accurate. The pulses are combined with other signals in the computer. This assures that all computer functions are coordinated at the proper time.

The memory of the computer is also on the system board. Memory is in two forms: read only memory (ROM) and read-and-write memory (RAM). The RAM memory was once called random access memory, thus the acronym RAM. ROM is permanent memory containing information that can not be changed by the

The integrated circuit chip holds information in memory in the form of electrical charges. These chips are quite small.

user. It is built into the computer. It is not lost when the computer is turned off. RAM is temporary memory that is lost each time the computer is turned off. It can be changed by the user.

The memory part of the system is in the form of electronic integrated circuits called chips. Each chip holds information in the form of binary digits or bits. These are encoded as electrical charges. The charges are stored in exact locations on the chip. This is similar to storing letters in certain boxes in the post office. Each box has an address for certain letters. An address decoder is included on the system board to keep the addresses organized.

Input and output ports are connected to the system board. These are used to connect the input and output (often referred to as "I/O") devices to the computer.

In some ways the computer is similar to a stereo music system. The amplifier is the heart of the stereo system. But, it is useless if there is no way to pick up the audio information from the record. There also must be some way to get the sound from the amplifier in a form that can be understood by the human ear. The stylus on the tone arm of the record turntable is the input device for the amplifier. The speakers are the output devices.

Input/Output Devices: Hardware

Information can be entered into the computer through several input devices. The most common of these include the keyboard, disk drive and joy stick. More sophisticated input devices include sensors such as those that monitor temperature in a laboratory experiment or air pressure in a wind tunnel. Common output devices include printers and video screens similar to television screens. Many early home computers were made so they could be connected to the home television set.

Nearly all the microcomputer systems will include a keyboard and a disk drive for input of information. A monitor is usually included as an output device so the work being done on the computer can be seen by the operator. Printers are very commonly used to provide a permanent, printed copy of the output. Before the computer is switched off, the work must be stored on a disk or printed on paper by the printer. If it is not saved, the information will be lost when the computer is switched off.

A keyboard looks much like a set of typewriter keys. When doing word processing, the keyboard serves the same function as the typewriter keys. The keyboard is the control panel of the computer. It allows the operator access to the computer system. Unlike a typewriter keyboard, most computer keyboards contain their own

Simplified Diagram of a Computer System

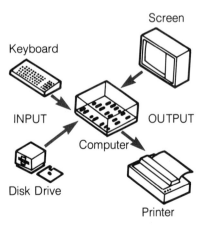

Computer Keyboard

A Keyboard Taken Apart.

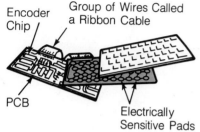

Encoder Chip
Group of Wires Called a Ribbon Cable
PCB
Electrically Sensitive Pads

electronic circuitry to allow them to communicate with the rest of the computer. The keys of the typewriter are connected directly to the elements which strike the paper.

Under the computer key pads there is a grid-work of wires. The wires run both horizontally and vertically much like screen wire. A key is located over each intersection of the wires. Electric current passes through the wires. When a key is pressed, the vertical and horizontal wires under it are pushed together. This way the microprocessor can detect which key was pressed. This information is sent to the CPU to be translated into a letter, number, or command. Several keys can be pressed at the same time to give different commands to the computer.

Keyboards can be programmed to convey special information to the computer. In a fast food restaurant, for example, the cash register may actually be a specialized computer. The keyboard is labeled with the names of each product sold: hamburger, large fries, small drink, etc. The sales person only has to punch in the order given by the customer (input). The computer is programmed to translate that information into prices (processing) which appear on the display (output). The name of each item and the price are printed on the sales slip that comes out of the machine (also output).

A joy stick is another device used to input data. It is often used with games but has other uses. The joystick allows the operator to move objects around on the screen more easily than could be done using the keyboard.

A joy stick is one type of controller used as an input device.

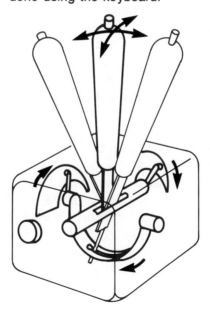

Computers in Technology

When a mouse is moved across a flat surface, a large ball inside turns two wheels. Counters record the movement. The information is passed on to the computer.

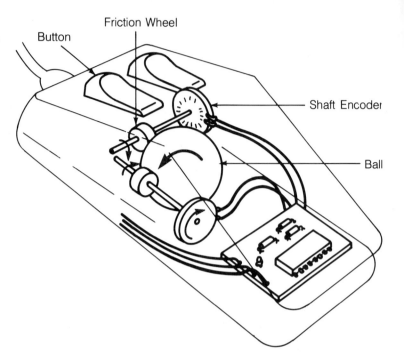

The mouse is another popular input device. As the mouse is moved over a flat surface, a roller ball underneath it turns. In one type of mouse, the ball turns two cylinders which are located at 90 degrees to each other. A light-emitting diode is placed beside a disc connected by a shaft to each cylinder. When the ball turns, the discs are turned. Slots cut in the discs cause pulses of light to flash through the discs. This light is detected by light sensing transistors. The signal is sent to the computer as information to be processed.

The magnetic disk is a very popular method of storing data in computers. It is made of a round piece of flexible plastic. The surface of the disk is coated so it can be magnetized. The disk surface is divided into parts similar to the slices of a pie. These are called sectors which are used to organize data. The data is recorded in circular tracks around the disk. In this way bits of data can be stored in a specific place along a track within a particular segment. This makes the data easy to find and retrieve. A directory (a sort of catalog) of the location names of the data is stored on the disk along with the data.

These disks are called "floppy disks" because they are flexible. They are sealed in a plastic wrapper for protection. There is a hole in the center of the drive shaft to turn it. A slot is cut in the wrap-

The disk and disk drive are commonly used input devices. The disk drive is also an output device. Information output from the computer can be stored on disks.

Exploring Technology

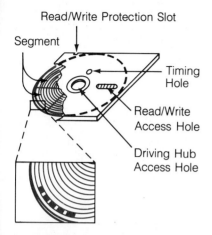

Magnified view of a floppy disk. Data is stored in circular tracks. The disk is also divided into pie-shaped segments.

Inside the Cathode Ray Tube Monitor

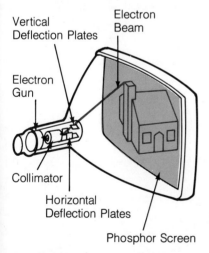

Most home computer systems include a monitor. The screen of the monitor is actually the front of the CRT tube, shown here from behind.

per to expose the disk to the recording head in the drive. The disk drive is designed to hold the disk in very accurate position. It also moves the recording head in a very accurate path in the recording area of the disk.

Other input devices include tape drives, graphics tablets, and light pens. The tape drive is similar to a cassette tape recorder. Information is recorded on the tape. Graphics tablets are used mostly by designers and artists. Light pens allow the operator to interact with images on the monitor.

The video monitor is the most commonly used output device. It looks like a television though it does not have all the electronic circuits needed in a television set. The screens of the monitor and the television are actually the front of a large electronic tube called a cathode ray tube (CRT).

In the CRT, an electron gun in the neck of the tube projects a beam of electrons to the front of the tube. This part of the tube is coated with phosphors that glow when struck by the electrons. As the beam of electrons flow toward the front of the tube, they pass through a "yoke" made of electromagnets. The magnetic forces from the yoke can be varied. This causes the stream of electrons to move rapidly left and right and up and down across the screen. The beam is also turned on and off when needed to form an image or leave a blank space in the phosphor coating on the screen. This forms a gridwork of dots of light and dark spots on the screen called pixels. (Pixels is a shortened version of "picture elements.") The computer controls the beam so that individual pixels can be illuminated when needed.

In monochrome (meaning "one color") monitors, one beam is switched on and off to form the image on the screen. In color monitors three beams form three images. Red, green and blue phosphors are excited by the beams of electrons. Varying intensities of these three colors can be created. This results in as many as 16 million hues of color.

Specialized CRT monitors can display as many as nine million pixels on the screen. Most monitors used on home computers can display only 128,000 pixels. With fewer pixels on the screen, sharpness is reduced. The cost of the monitor is also less.

The screens on televisions and computer monitors are not flat. They are slightly curved. The neck of the tube also projects from the back of the monitor. This makes the monitor a bit bulky. The demand for portable computers has led to the development of flat-panel monitors or displays. These are made with grids of thousands of tiny electronic elements. The elements are arranged in

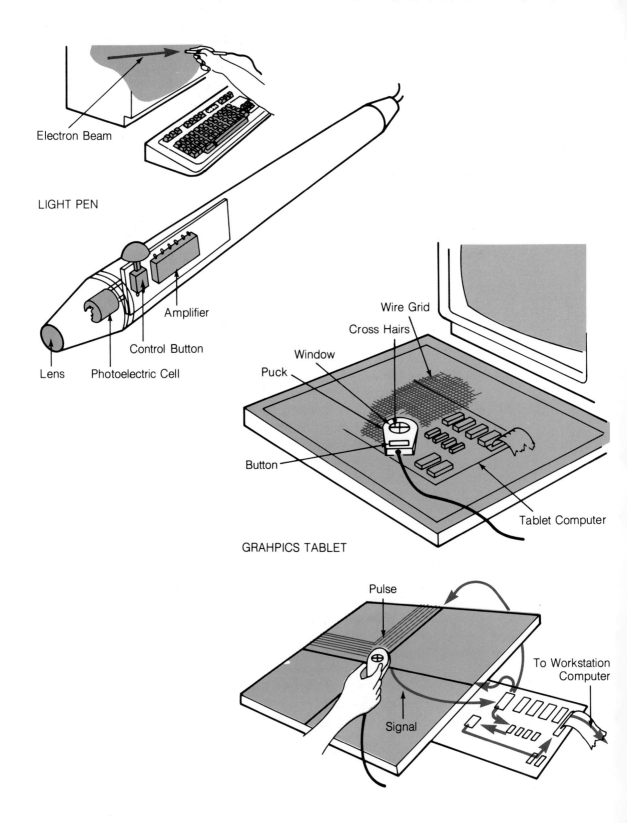

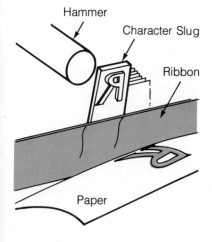

Daisy Wheel Printer Head

The basic elements of an impact character printer are a hammer, a character slub, an inked ribbon and paper. A hammer stroke pushes these elements together to print the character on the page. In models that print a single character at a time, character slugs are commonly arranged on the perimeter of a circular element called a daisy wheel; the hammer strikes a slug, which is brought into position by highspeed revolutions of the wheel. In some line printers, the character slugs for a line are arranged on a flexible metal band and the paper is then struck against each character.

Daisy Wheel

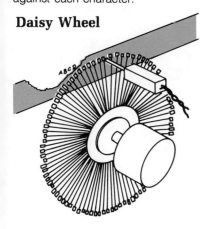

rows and columns similar to the openings in screen wire. Each opening can be made to darken or emit light when electricity is applied. This is controlled by the binary on-off instructions from the computer. Generally, flat-panel displays do not have the sharpness of the CRT displays. This is an area of continuing research.

Printers provide a permanent paper copy of the computer output. Impact printers form the printed image by pressing an inked ribbon against the paper. This is very similar to the action of a typewriter. Non-impact printers form an image in a variety of ways.

There are two major methods for forming impact images in computer printers: dot matrix and "daisywheel." Daisywheel refers to the fact that the printing element in the machine is made up of many petal-like spokes around a center shaft. Each element contains one letter, number of symbol. The wheel turns to place the proper character to the top, over the paper. The hammer-like rod strikes the character against the inked ribbon which is pressed against the paper. This action is done rapidly as the wheel spins, bringing each character into the proper striking position.

Daisywheel printers produce a sharp image, similar to the letter-quality typewriter images. They are limited in the number of different charcters that can be printed. Though the wheels can be changed in most machines, the printer still can print only the characters included on a wheel. Daisy wheel printers cannot print graphics. This is a major limitation of this kind of printer.

Dot matrix printers are more versatile than other types of printers. The name " dot matrix" comes from the fact that the letters are formed from a pattern of small dots. They can print graphics and a large variety of characters. Tiny rods or pins are arranged in vertical columns in the print head. There may be several columns of pins in the print head. Each pin acts as a hammer and can be controlled independently. Each pin can be made to strike the ribbon, pressing it against the paper. This forms a dot of ink on the paper. As the print head moves across the paper the pins are fired hundreds of times. Different combinations of the pins can be used to form a wide variety of letters, numbers and symbols.

Dot matrix printers are very popular because the number of characters that can be formed is almost unlimited. The image, because it is made up of dots, is not as sharp as images produced by the impact printers. This is a disadvantage when sharp, letter-quality images are needed.

Non-impact printers include ink-jet, laser and photo methods of forming an image. Ink jet printers form images made up of dots

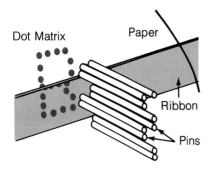

Here, one column of nine pins in the head of a dot matrix printer has formed the letter "B."

Credit: Apple

similar to the dot-matrix printer. Tiny droplets of ink are shot or sprayed onto the paper to form the characters. As the print head moves across the paper the computer-controlled nozzles are activated.

Color printing is possible with ink-jet printers. Several ink jets can be put in the print head for the basic colors, cyan, magenta, yellow and black. The droplets of colored ink can be shot onto the paper in varying patterns to produce colored images.

Laser printers produce a very sharp image similar to printed copy. This printer is an outgrowth of electrostatic printing methods used in copy machines. A small laser, pulsing millions of times per second, creates an image on a polished metal print drum. The drum is charged with a positive electrical charge. The laser light beam neutralizes spots on the drum in the shape of the image

The dot matrix printer is a popular output device.

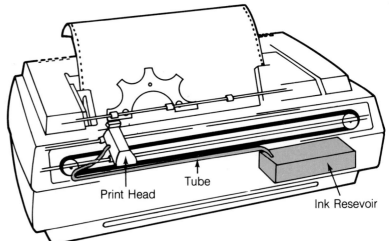

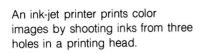

An ink-jet printer prints color images by shooting inks from three holes in a printing head.

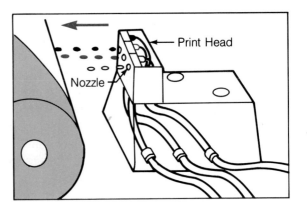

Exploring Technology

A laser printer produces typeset-quality text. Graphics can be printed similar to original art work.

to be printed. A printing powder is dusted onto the drum where it sticks only on the neutralized areas. When the paper runs over the drum, the powder sticks to the paper forming the printed image. Heat and pressure are used to fuse the image onto the paper.

Plotters are used to produce large graphics such as drawings by designers and architects. A plotter is usually used to output images created by CAD programs. Most plotters have a set of ink pens which are moved over the paper to create a drawing. Colored pens can be used to add color to a graphic. Many plotters will automatically change pens as needed.

Software

Software is a computer term meaning the lists of instructions, called programs, that tell a computer what to do and when to do it. The term has been widely used since about 1960. It is used to separate the instructions that control a computer from the computer itself. Computer programs are stored as magnetic symbols on disks or tape. The programs are read into memory in the computer from disk or tape drives.

There are two main types of computer programs: operating systems and application systems. The operating systems programs tell the computer how to interpret instructions and data. They also tell the computer how to control the other machines attached to it. These include the disk or tape drives, the printer, and the modem. Operating systems programs are the various programs used to manipulate data. They include word processors, spread sheets, data bases and games.

Programs must be put into computer memory before they can be used. When a computer is turned on, built-in programs in the ROM give the CPU start-up instructions. The ROM is told how to find the operating system on the disk or tape. It is told to copy this information into RAM, the temporary memory. At this point the operating system program begins giving instructions to the computer. User commands are set up so the operator can control the computer. This is referred to as "booting" the computer. An application program is then read into RAM. This program tells the computer what to do with the information about to be entered.

An example may help better explain the process. Assume that you are about to write a paper for a homework assignment. You decide to write it on a computer, using it as a word processor.

Before turning on the computer, you put a disk containing an operating program into a disk drive. In the other disk drive you place a blank disk. When you turn on the computer the first drive begins running and words appear on the monitor. At this point you are "booting" the system.

Usually some instructions appear on the screen telling you to do certain things such as enter a new date or time. This gives the computer information it needs to continue the process.

The first disk you put in contained the operating system. This, is put into memory along with the program in the permanent memory, ROM, to tell the computer what to do. You then put the second disk, the word processing program in the first disk drive. This program tells the computer what to do with the information you are going to enter through the keyboard.

As you begin entering your paper by typing on the keyboard, both programs are being used by the computer. You can see this by watching the indicator lights flash on and off as each drive runs. As you type in the information (your paper), the word processing program tells the computer what to do with the information. Lines

Computers are becoming commonplace in schools. Computer literacy is a concern as the computer is used more and more in our lives.

Sample print-out of a software program. This program was written in the computer language BASIC.

```
30230 LOCATE 5,13:PRINT AC;:LOCATE 5,70:PRINT TN$:LOCATE 7,28:PRINT YM$:LOCATE 9
,22:PRINT SN$:LOCATE 11,24:PRINT USING"##,### lbs.";W!
30240 LOCATE 13,25:PRINT PD$;:LOCATE 13,68:PRINT USING"$$##,###.##";PP!:LOCATE 1
5,37:PRINT USING"###,###";OO!:LOCATE 15,72:PRINT USING "###,###";HO!:LOCATE 17,2
8:PRINT A$(DM):LOCATE 17,62:PRINT DP;:LOCATE 19,42:PRINT O2$
30360 P1=15:PI=37:O3!=OO!:O!=0:GOSUB 4300:IF A$=CHR$(27)THEN RETURN ELSE OO!=VAL
(AN$):O3!=OO!:RETURN
32570 FOR I=1 TO K1-1-D:FOR J=I TO 1 STEP -D:IF C#(J)<=C#(J+D)THEN 32580 ELSE SW
AP C#(J),C#(J+D):NEXT J
32580 NEXT I:D=INT(D/2):IF D>0 THEN 32570
32590 FOR I=1 TO K1-1:P(I)=INT((C#(I)-INT(C#(I)))*1000+.5):C!(I)=INT(C#(I))/1000
:NEXT I:ON DO GOTO 32600,32700
32600 GOSUB 7000:LOCATE 4,1:PRINT U1(T0):IF T0=1 THEN PRINT SG$; ELSE PRINT SPAC
E$(23);"0--------10--------20--------30--------40--------50---55+";
33100 LOCATE 4,5:PRINT U1(1):PRINT SG$;:GOSUB 34000:IF A$=CHR$(27)THEN 33225 ELS
E 33210
33622 IF C=>K1 THEN LPRINT"":LPRINT STRING$(79,61):LPRINT"":LPRINT PC$(1);PC$(3)
;"Total number of vehicles ==> ";K1-1;PC$(2);PC$(4):CP=CP+4:GOSUB 32740:GOTO 336
25
33624 IF CP>59 THEN GOSUB 32740:PC=PC+1:GOTO 33610 ELSE 33620
```

and paragraphs are formed. Titles and subtitles are formed and centered on the page as needed. Lines are indented. Special words are put in italics or bold type or are underlined. All these things are done by the word processing program from commands you give it by pressing keys on the keyboard.

When you are ready to print your paper, the proper command is given. This tells the operating program to send the information to the printer. The operating program in the computer is controlling the printer.

Types of Software

You could have done many different things on a computer instead of writing a paper. Accountants use spread sheets to keep track of financial matters. Data base programs are used work with large amounts of information such as lists and research data. Games are examples of special software programs.

Word processing programs are used to write and edit letters, research proposals, and papers such as the homework assignment used in the above example. The authors of this book used word processing programs to prepare the manuscript to be sent to the publisher.

Word processing programs allow the writer to enter, change, move, replace or delete parts of a document being written. A spelling checker program is available for many word processors. This special program helps eliminate spelling errors. Some spelling checkers contain a dictionary of thousands of words. Usually new words or special abbreviations can be added to the dictionary. Word processing programs for computers are making the typewriter obsolete. It is very easy to edit and rearrange thoughts or correct mistakes before printing.

When the material is composed and all corrections have been made, a copy is printed. A file copy is not needed if it has been

Sample print-out of a spread sheet.

```
MONTHLY OPERATIONS REPORT    SEWER CITY WASTEWATER TREATMENT PLANT
==================================================================
Day   Totalizer        Flow Rates        Precip      Temperature
      Reading                                            Air
                     Avg       Max                   Max   Min   RBC
      1000 gal       gpd       gpd       inches     degrees C
------------------------------------------------------------------
  1   23456789      51000    110000       0.00       29    18    22
  2   23456840      58000    120000       0.40       29    21    22
  3   23456898      52000    130000       0.05       26    20    23
  4   23456950      54000    135000       0.00       26    15    21
  5   23457004      40000    120000       0.00       30    24    22
  6   23457044      51000    110000       0.00       32    13    22
  7   23457095      34000    100000       0.00       32    20    22
  8   23457129      59000    130000       0.00       32    15    22
  9   23457188      55000    125000       0.00       33    13    23
 10   23457243      63000    150000       0.20       32    22    23
 11   23457306      88000    190000       0.90       33    18    23
 12   23457394      56000    140000       0.00       33    18    23
 13   23457450      57000    140000       0.00       32    18    23
 14   23457507      52000    120000       0.00       31    18    23
 15   23457559      58000    150000       0.00       28    19    24
 16   23457617      59000    140000       0.00       23    17    23
 17   23457676      56000    140000       0.00       31    21    23
 18   23457732      70000    150000       0.00       33    19    23
 19   23457802      63000    150000       0.00       35    25    23
 20   23457865      54000    160000       0.00       34    18    23
 21   23457919      43000    130000       0.00       32    17    23
 22   23457962      58000    160000       0.00       33    17    23
 23   23458020      61000    130000       0.20       31    21    24
 24   23458081      62000    130000       0.00       31    21    23
 25   23458143      61000    140000       0.00       28    21    23
 26   23458204      60000    140000       0.30       30    17    23
 27   23458264      76000    120000       0.70       31    21    23
 28   23458340      55000    100000       0.00       32    19    23
 29   23458395      59000    140000       0.00       32    23    23
 30   23458454      59000    120000       0.00       28    23    23
 31   23458513      57000    130000

TOTAL
```

saved on the disk. Some authors write using a computer then send their disks to the publisher. The disks can be put directly into typesetting machines. In this way it is possible to eliminate all paper copies before a book is printed.

Spread sheets are used by bookkeepers and accountants to keep track of financial matters. A computer spread sheet program divides the page on the screen into columns across the page and rows down the page. Each row and column is labeled with a number or a letter. This creates a checkerboard effect of squares into which information can be placed. This allows the computer to keep track of the information.

Spread sheet programs are written so they can do mathematical calculations needed. The math is performed using formulas which tell the computer what to do with the numbers in the squares. For example, the formula sum (a1+ a2 + b4) tells the computer to add the contents of squares a1, a2 and b4 together.

Spread sheets have a variety of uses. They can be used in the home for keeping track of checkbook balances or household accounts. Today businesses must use computer spread sheets to keep track of their complex bookkeeping. Computer spread sheets are a good tool for predicting future financial conditions. Formulas can be changed to look at the effect changes would have on a business. A company could look at the effect pay increases would

An engineer uses a light pen while designing an automobile part

have on the profit for a year. What would an increase or decrease in the number of employees do to total production of a product? How would a change in the percentage rate of interest on loans affect the monthly payments for a mortgage loan? Answers to questions such as these could be found before the changes were made.

Spread sheets are being used by many people to calculate income taxes. Each month new information is added. At the end of the year all the information needed to complete the tax return is available. People who prepare tax returns have special spread sheet programs which print the required information directly onto the proper forms. What once took hours or days to record and calculate can be done in minutes.

Data base programs are used to organize large amounts of data. A data base program can be compared to a filing cabinet. The information can be organized into categories called drawers, files and documents. It can be moved around in the data base and organized in a number of ways.

Here is an example. The list of thirty customers on your paper route includes names, addresses, and telephone numbers, where to leave the paper, and the cost of the paper each week. Each month your work piles up. If you had this information in a data base you could simplify the work. With a little planning you could easily go to the computer for a customer's bill, or for a list of customers. The list could be organized in alphabetic order, the order the papers are delivered, or a list of those who have not paid their bill. As a good business person you might want to send seasonal greeting cards to all your customers. Your data base program could be used to print out mailing labels for the envelopes.

This example lets you see how information can be rearranged to give new information. The reports made from a data base give you power over information. Businesses use this program to record inventory. Auto parts stores use data base programs to keep track of the thousands of different items they must handle. Many large grocery stores and department stores use data base programs in computers connected to the registers at the check-out counters. A scanner in the counter reads a number from the label on the item. This tells the computer that the item has been sold and to remove it from inventory. Records of the sales can be used to print lists of items that need to be reordered as well as lists of all items sold. A complete inventory of all stock remaining in the store can be printed very quickly.

Setting up a data base starts with the list of information you want to work with. Data about the paper route would be entered in a form similar to the example below.

NAME: Jones, John
STREET: 165 S. First St.
TOWN: Anywhere
STATE: VA
ZIP CODE: 00000
PHONE: 000 0000
PAYMENT: $14.25

Notice that each piece of information is listed separately. No items can be combined. This allows you to organize the information in any way desired. Adding and correcting information is easy. Customers can be added or removed from the list and changes can be made in existing records. Up-dating information is very important when keeping records.

Games are an example of special computer programs. Playing games on a computer can be exciting. Their popularity has made household names out of many of these games. Game characters seem to come to life on the computer screen. They appear in colors and even change colors when "zapped" or captured.

The computer game makes use of graphics programs to create the figures. The movement of the character on the screen is done with a controller such as a joy stick. The computer game is programmed to be interactive. Input from movements of the joy stick tell the computer to move the character, make sounds and change the score.

Computers Today

Computers are used in a wide variety of ways. Hardly any area of modern life is untouched by the computer or the technology that makes it possible. In earlier parts of this chapter several uses of the computer were shown in business, the home and in industry. In other chapters the computer has been shown to be a vital tool in all systems of technology. The application of computers are limited only by human imagination.

In education, for example, the computer is having great impact. Can you remember when the first computer was put in your school? How many times and in how many ways are you using a computer in your school work? Are you using word processors when you write school papers? Are computers being used as tutors to help students? They can make learning easier. They also can help you learn more.

Computers are finding much use in agriculture. Dairy farmers, for example, can computerize their operation. A small radio unit can be put around a cow's neck. When she enters the milking parlor, a signal is sent to a computerized feed dispenser. The cow receives the correct amount and mixture of feed designed to give her the nutrition she needs. When the milking is done, the amount of milk collected is recorded. Complete records can be kept on each cow to determine the optimum output of milk.

The dairy farmer also can use the computer to inventory farm supplies and equipment. Maintenance records on each machine can be kept to determine the best time to replace the machine. Records of personal and farm finances can be kept easily using a spread sheet program.

Industries are finding that they can not operate without computers. Engineers design parts on a computer.

They then program stresses to which the part will be subjected. Weak points show in color on the computer screen. The part can be redesigned until all weak spots are eliminated or minimized.

The doctor's office is now being computerized. Each patient's records can be entered into computer memory. A complete

Touch typing skills can be taught with this educational "toy". It evaluates progress and scores each lesson for accuracy and speed.

Designers use the computer to design and test products before they are made.

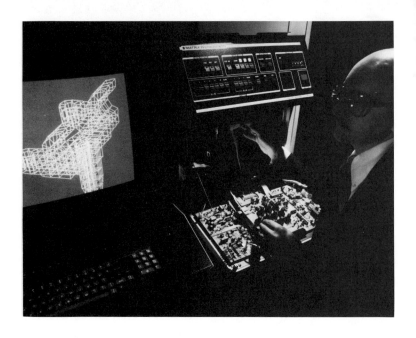

medical history can be developed. From the computer records the doctor can look for trends which indicate possible future problems. Of course, all financial records can be kept on the computer.

These few examples, along with the illustrations throughout this book, show the impact of the computer. Our lives are dependent on this marvel of electronics. Our cars run because of a computer under the hood. When something goes wrong with the car, a computerized analyzer is used to pinpoint the problem. The grocery store could not sell its products if the computer failed to control the cash register. Use of computer technology has made many wonderful things possible. As we become more dependent on these things, we are becoming more dependent on the computer.

Suggestions For Further Study

Observations

Computers are all around us. Can you find computers or examples of computer technology being used in your home? Are microcircuits used in your home appliances? What about the family automobile? Microcircuits are used in automated cameras. Look for examples of the use of computer technology in your every-day life.

Do you have a computer in your home? If you do, then you probably already know a lot about computers. How does your home computer compare with the ones used in your school? How does it compare with the computers used in businesses and industries in your community? Your teacher may be able to arrange a trip to see some of these computers.

Can you write a computer program? Using the BASIC computer language, you should be able to write a simple program. The program listed in the activity at the end of this chapter will give you experience in entering a program and running it. Many computer magazines print easy-to-use programs which can be entered and used.

Questions

1. Briefly trace the history of computers from the abacus to modern times.
2. What was the feature that separated first generation computers from all others? Second generation? Third generation?
3. Has a fourth generation of computers started? Why?
4. What is a computer?
5. What are the main parts of a computer system?
6. What is the difference between RAM and ROM in computers?
7. What are the two types of computer programs?
8. Describe the different types of computer software. How is each used?

Key Words

computer	terminal	light pen
integrated circuit	CPU	pixel
microcomputer	ROM	dot matrix
abacus	RAM	"daisywheel"
calculator	BASIC	letter-quality
relay	keyboard	ink-jet
electronic tube	joy stick	software
transistor	"mouse"	data base
core memory	"floppy" disk	spread sheet
"main frame"	graphics tablet	word processing

Computer Programming

Concept: *Programming*

Computers must be told what to do. These instructions are called programs. Without programming, a computer is nothing more than a collection of parts.

The computer may have had the greatest impact on the history of humans of any machine ever invented. Never before has a machine generated so much interest in such a short time. Today it is difficult to find an area that has not been touched in some way by computers.

We are rapidly approaching the time when a literate person will have to be literate in computers. You also need to learn to use computers to do as many things as possible.

Objective

This activity will help you get started using computers in a simple, easy way. As you learn more about computers you will want to move on to the larger activities in the Activity Manual accompanying this text.

These simple programs are written in the computer language called Beginners All-Purpose Instruction Code (BASIC). BASIC is a very popular language. It is a mixture of words, numbers and symbols. It is easy to learn, yet has many uses. There are many forms or "dialects" of BASIC. It is important for you to understand the dialect for the computer you are using.

Procedure

To use the programs listed here, follow the instructions given below. The instructions are written to explain the use of the simple temperature conversion program. They are written for use on an IBM computer system.

1. If your computer requires the loading of a disk operating system (DOS), put the disk containing the DOS in the disk drive of the computer. In computers with two disk drives this will be the top or left hand drive. (This is usually referred to as "drive A".)
2. Turn on the power switch. The computer will run for a few seconds. It is checking its memory and all attached hardware to be sure everything is working properly. The system beeps when everything is ready.
3. The first thing you see on the screen is the date message. To simplify this activity, merely press the ENTER key. You will see a time message. Press ENTER again. You could have entered the current date and time if needed.
4. A start-up message may be displayed.
5. A "prompt" will be displayed. On IBM computers this will be A> followd by a blinking line. This blinking line is the cursor which shows where the computer will display whatever you type next.
6. The computer is now waiting for a command. Type in BASICA and press RETURN. This tells the computer that you want to enter a program in the BASIC language. A message about BASIC will be displayed. The blinking cursor will be below the message. The computer is now ready to accept your program.
7. Type in the temperature conversion program one line at a time. At the end of each line press the

ENTER key. This marks the end of a line and enters the command into memory.

TEMPERATURE CONVERSION

```
10 LET C = 20
20 PRINT "THE CELSIUS TEMPERATURE IS";C
30 LET F = (9*C)/5 + 32
40 PRINT "THE FAHRENHEIT TEMPERATURE IS";F
50 END
```

8. When all lines have been typed in, type RUN. The computer will print a message on the screen. Notice that this message includes the solution to the formula in line 30.

To convert any celsius temperature to fahrenheit, type the celsius reading into line 10 in place of the figure given there. Type ENTER at the end of line 10 after you have typed in the new figure. Move the cursor to the end of the program. When you type RUN, the computer will calculate the new fahrenheit temperature. A message will be printed on the screen. Try calculating the fahrenheit equivalents for several celsius temperatures.

Chapter 10　Consequences of Technology

Introduction

In Chapter 1 you learned the meaning of technology. You also learned about the major systems of technology: production, communication, and transportation. You learned that technology is all around you. Even the trees in a nearby forest may be products of technology.

You have learned that technology causes change. Things change because of technology. You also learned that technology has an effect on each of us. Now that you have explored technology in some depth, you should be able to better understand the model shown on page 270.

The model shows ways that people and technology interact. Something is not obvious in the diagram, though. It does not show the opposite effects on humans. There are forces that oppose the way people interact with technology.

Consider this example. In a football game there are two teams. One is the offense, the other the defense. The defense could be called an opposing force. It opposes or tries to stop the offense. A similar thing occurs in technology. For each force in technology there is an opposing force. Let's look at several parts from the model. We can look at some forces and opposing forces or *counterforces*.

Using technology to recycle our limited resources.

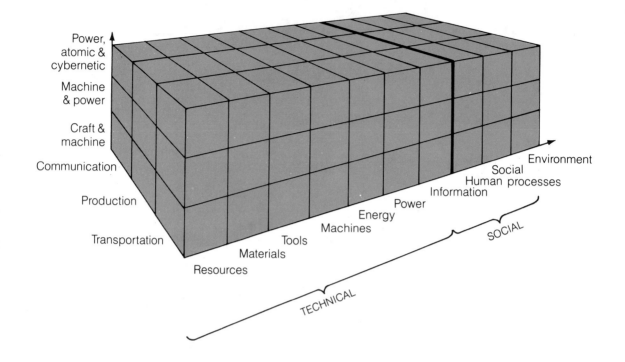

Eight Counterforces of Technology

Materials

In Chapter 1 you learned that many materials are used to make our lives easier. Modern steel, plastics, and synthetics are examples. These materials are a part of our material culture.

So, what is the counterforce? It is, "Nature controls materials supply." This says that there is a limit on the supply of materials. We know this is true. The limited amount of petroleum is an example. Since Drake's first oil well, people have become dependent on oil. Many uses have been found for petroleum products. But, we also are beginning to realize that there is a limit to the amount of oil in the earth.

We enjoy using many natural materials. But many of them are limited. Nature controls the supply of them. Nonrenewable resources (see Chapter 1) are in limited supply. We can supply ourselves with the renewable resources.

Nature

Technology helps people gain advantage over nature. When people gain advantage over nature they overcome the forces of nature in many ways. Houses protect us from weather. We have learned to make an airplane that will fly through the air, much like a bird. In both examples people have overcome the forces of nature.

What is the counterforce? Nature retains authority over man. What does this mean? Consider the airplane again. What would happen if an airplane wing was made like a square box? Would it provide the lift needed to make the airplane leave the ground? Not at all! We know that the needed lift will not be provided by this wing. This is a limitation of nature. Through much research, it has been found that an airplane wing of a certain shape will provide lift. The many experiments with shapes of airplane wings showed many that would not work. It also showed some that would. The researchers had to overcome limitations of nature. They had to find what nature would let them do before the airplane would fly.

This is the meaning of the counterforce, "Nature retains authority over mankind." Through science and technology, people have found ways of working *with* nature. But, nature retains the upper hand. People have to work within the limits of nature. If an airplane is built with square wings, it will not fly.

Think about this for a minute. Can you think of other examples of the limitations of nature?

Environment

People create their own environment. Throughout this book you have seen examples of this. Houses, factories, bridges, and skyscraper office buildings are constructed. Transportation systems carry us to faraway places. The highways, airports, train tracks, and waterways affect our environment.

The counterforce of this states that the environment has an impact on mankind. We have not been able to control the weather. Hurricanes and tornadoes still cause millions of dollars of damage each year.

The environment we create also affects us. We have created a communication system through television. The television affects us. We spend time watching it instead of working at creative hobbies. Some people feel that television affects reading habits of children.

Through our technology we have changed the natural environment. Air pollution grows so severe in some areas that people become ill. Fish are killed by polluted water in rivers and lakes. These examples show vividly that environment has an impact on people.

Acid Rain. One effect of air pollution has become increasingly apparent during the last several years. "Acid rain" has killed fish and other aquatic animals in hundreds of lakes in the United States. This has been caused by an increased acidity of the water.

The acid comes from precipitation (rain or snow) which contains sulfuric and nitric acids. These acids form in clouds when sulfur oxides and nitrous oxides interact chemically with sunlight, moisture and oxygen. Generating plants, industries and motor vehicles give off sulfur and nitrous oxides as a by-product of burning fossil fuels.

The concentration of acid rain is very high in the Northeast, particulary upper New York and New England. The Ohio Valley Region has the highest concentration of acid precipitation. This is also an area of highly concentrated industries.

The high acid content of lakes kills fish. Trees and other plants are damaged. Acid rain (or fog) can cause stone and steel buildings to slowly deteriorate. Stone buildings and sculptures in historic Greece which have stood for centuries are now cracking and eroding from sulfur oxides in the atmosphere.

Though many efforts are being taken to stop this serious environmental problem, the answers are not simple or easy.

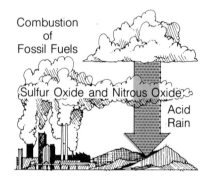

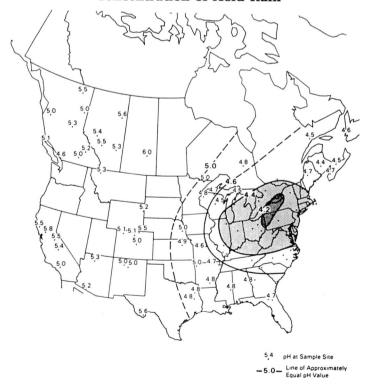

Concentration of Acid Rain

Culture

Our culture has developed throughout our history. Our music, art, tools, and machines all are a part of our culture. Technology is a part of this culture.

Another counterforce is that culture shapes its own maker. One example of this is the effect that change has on people.

What would happen if a modern Rip Van Winkle woke today after forty years of sleep? What would he see? How would it affect him?

Without a doubt he would be shocked. He would see forty years of change all at once. He would not have become accustomed to the changes gradually as they happened. He would feel the impact of the changes suddenly.

Sudden or fast change is hard to accept. Some people cannot cope with sudden change. Alvin Toffler, in his book *Future Shock*, talked about this reaction to sudden change. He talked about shock caused by too rapid movement into the future. Much of this change is caused by technology.

How fast has change taken place in your life? Change affects all of us. Ask your parents how their childhood differed from yours.

Leisure

You learned in Chapter 1 several ways that technology affects leisure time. Books made reading a leisure time activity available to anyone. Most modern hobbies are possible because of our level of technology.

The counterforce of this is that our leisure activities affect our culture. As technology developed, people had to work less and less. Work became easier. Less human effort was needed. Less time was required. People worked fewer hours each week. People were able to spend more time with leisure activities. People could spend more time on activites such as art, music, and reading.

People, who once had to spend all their time working, now can spend time learning about such things as art and music. This changes their lives. Hobbies and leisure time activities are restful and relaxing. People need time to relax and rest from the workday.

Change in Humans

Human beings created themselves through technology. Most people are healthier today than their grandparents were. We have better medical care. We have better ways of treating illnesses. The average life span is increasing.

This crop duster is spreading an insecticide which may protect a crop from insects, but which may also harm humans.

The counterforce here says that there is a danger in this. It is possible to go too far with our technological innovations. Scientists have harnessed the power of the atom. Atomic energy can provide electricity to light and heat our homes, offices, and industries. It also can kill millions of people when made into a bomb.

Nuclear radiation can help cure diseases. It can be used in agriculture to increase food supply. Radiation escaping from a nuclear power plant could kill millions of people. Insecticides are a product of our technology. They can be used to kill insects that eat our food crops. Insecticides can be used to kill insects that carry diseases people can catch.

These same insecticides also can harm humans. Workers in a chemical plant making insecticides may become ill. Dust from the insecticides being made can do permanent damage to them. If used improperly, the insecticides can kill helpful insects also. Helpful bees can be killed as well as the harmful beetles.

Our technology is helpful in many ways. It also can be harmful to the very people who created it. Today people have the ability to destroy themselves.

Expression

Throughout this book you have seen ways people can express themselves. Architects can express themselves in the buildings they design. Artists can work with plastic, steel, and other modern materials of technology. People use technology in expressing themselves.

The counterforce of this is that technology limits this self-expression. We can depend on modern conveniences so much

In large vineyards, grapes may be picked by machine instead of by hand.

that we overlook older ways of doing things. A woodworker has many machines to use. It is easy to forget the hand tools. Some jobs can be done easily, even better with the hand tools. However, the worker may be so used to machines that the work is done the hard way with the machines. The woodworker's self-expression has been limited.

Work

Several times in this book it has been mentioned that the work week has been shortened by technology. People do not have to work as long today as they did 50 or 100 years ago. Work now requires less human effort. We know that our changing technology constantly creates new areas of work for people.

What is the counterforce to this? Technology also eliminates and changes work. Although new jobs are created, old ones are eliminated. You can think of many jobs that no longer exist, or at least, are now done by fewer people.

How many people do you know who make buggy whips? Do you know what a buggy whip is? How many people now make horse-drawn buggies? In the days of horse-drawn transportation, the buggy was an important vehicle. There was a whip on each buggy for the driver to use to hit the horse to keep it moving.

How about the village blacksmith? Are there many blacksmiths working today? A few are needed to shoe horses on farms or at race tracks. Today blacksmithing is almost a lost art. Where it is done, it is mostly a craft area. Today people put high value on handcrafted metalwork done by a blacksmith.

It has been mentioned many times that change is a part of technological progress. This is true with work also. The brief examples given above show change taking place in work. They also show change taking place by the creation of new jobs and the elimination of old jobs. In many ways the most noticeable changes take place in the area of work. When a person's work is no longer needed, that person may feel very bad about the loss. To have to change jobs can be upsetting. Changes in work cause changes in a person's everyday life.

Systems of Technology

Another way to look at the consequences of technology is to look at the various systems in technology. You know that technology can be divided into three major systems: production, transportation, and communication. In this book, these have been divided further. We have used processing, energy and power, manufacturing, construction, transportation, and communication as our major chapter headings. So, let's look at the consequences of technology in each of these systems.

Transportation

Research and travel into space have brought about many consequences in travel. Airplane pilots had problems with airplanes skidding on wet runways after a rain. It was discovered that grooves in the pavement allowed the water to run off. This gave better contact between the tire and the pavement. Grooved pavements are now used on many highways. Wet pavement accidents have been reduced.

Guard rails and barriers on highways can be shock absorbing. The barriers can be made so they absorb the impact of an automobile traveling 88 kph (60 mph). These same shock absorbers can be built into the bumpers of cars. This use of a development made in space research has made highway travel safer. Many more examples could be given. How many can you find?

Many consequences of transportation technology have been negative. Many problems occur. On the December 14, 1968 cover

Helicopters transport people to hospitals to provide fast emergency medical care.

of The Saturday Evening Post, the question was asked, "Are We Heading Toward the Day Everything Stops?" On the cover was an artist's drawing of examples of advances in transportation technology. Automobiles and traffic jams were shown. Ships, airplanes, and trains were shown with their own kinds of problems. Above it all the sun looked down with a sad look on its face.

Is the day coming when everything stops? Is this possible? Is it possible that all our transportation will stop? Is the automobile the transportation vehicle of the future? As more and more airplanes are put into the air, will the skies become overcrowded? Will airports be overcrowded? Can our system of air traffic control be adequate?

The predictors of the future say automated highways are possible. Trains that travel 800 kilometers (500 miles) per hour are possible. Some people say we now have the technology to develop these things. We have the technical ability to make automated highways and high-speed trains that work well.

Most of you reading this book will live most of your lives in the twenty-first century. What will life then be like? What will transportation be like? How will people move from place to place? How many people will have to be moved? What will happen if petroleum supplies are gone? How will internal combustion engines be fueled? Will there be any internal combustion engines?

Consequences of Technology

Someday you may ride to work in an electrobus.

Questions such as these must be asked. They must be considered by technologists. We can't wait until a resource is gone, then think about a replacement for it. First, we must think of ways to save as much as possible now. We must eliminate waste. We must also be working to develop other materials that will do the same job.

As was mentioned in Chapter 6, Transportation, we have become dependent on the automobile. Our way of life depends on

How will transportation vehicles be designed in the future? Here is an artist's idea of the type of work station designers will use in the future.

336　　　　　　　　　　　　　　　　　　　　　　　　　　　Exploring Technology

an automobile to carry us to many places. Large numbers of people in cities need automobiles to transport them to work and home again. Over the years people have gradually moved farther and farther from their work because they had an automobile in which to travel.

Mass transportation systems have not developed. Our governments have spent millions of dollars in highway construction. The automobile industry has spent millions of dollars developing automobiles.

What would have happened if some of this money had been spent on developing mass transportation systems? We are beginning to realize that mass transportation is necessary. One train car can carry as many people as ten or fifteen automobiles. The one train car causes little air pollution. Fifteen automobiles produce a lot of air pollution.

But a change from personal automobiles to mass transportation will not be easy. Technically it can be done. But, technical problems are not the biggest problems. People are accustomed to traveling in their personal autos. They like to come and go when they want. People do not like to change. The problem is really a social problem. People's attitudes must be changed. The technical problems can be solved. Can we solve the social problems?

Energy and Power

You have seen in Chapter 3 some of the consequences of energy and power technology. We now live a life in which people have to do little physical labor. Many advances in technology have made life easier for us. The human muscles are no longer a major source of energy and power.

But we have developed an energy-intensive way of life. This means that many of the things we do require a lot of energy. Our automobiles require large amounts of petroleum. Our homes require large amounts of fuel to heat them in the winter. Many modern buildings are made so that they must have air conditioning in the summer. Windows can't be opened to admit cool air. Instead, machines move cool air through the building. It requires large amounts of electrical energy to run the air conditioners in these buildings.

Our biggest source of energy has been petroleum. Now we are finding that we are running short on oil. Certainly oil is a finite resource. There is a limit to the amount of it in the earth. No more is being made. It is a nonrenewable resource. In the past we didn't think seriously about this. Oil was not expensive. There seemed to be plenty of it to use. We now realize that we are dependent on the oil, now that it seems to be in short supply.

This is a negative consequence of energy and power technology. There are also positive consequences. Through our technology it is possible to get energy from many sources. There is much interest in solar, geothermal, tidal movements, and nuclear sources. In varying amounts, these sources offer hope for the future. Each source has problems that must be overcome through new technology.

Solar energy is plentiful. As long as the sun is shining, there is plenty of energy available. But, that is part of the problem. When the sun is not shining, there is no energy available. So, ways of storing the energy must be developed. More efficient ways of collecting the solar energy also must be developed. We collect only a small part of the total energy available.

Geothermal energy is available in only a few places on earth. Theoretically it is possible to get geothermal energy from any place on the earth. All one has to do is to drill to the hot center of the earth and use the natural heat there. But, the technology to do this has not yet developed.

Tidal power has similar problems. Few places on the earth have tides large enough to produce enough energy to be worthwhile.

Oil is a nonrenewable resource which is becoming harder to find. This rig extracts oil from deposits beneath the ocean floor.

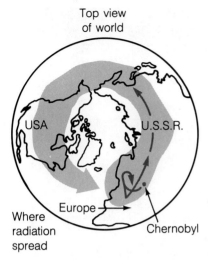

Nuclear energy has a well known problem, a large radiation hazard. The nuclear power plant can produce large amounts of electrical energy. But, there is the danger of an accident that will release radiation. Most power plants are built in the areas where the energy is needed. This is in the large population centers. This means that radiation released from one of these power plants would affect large numbers of people.

A disastrous accident did occur on April 26, 1986, at the Chernobyl nuclear power station in the Soviet Union. Thirty-one people died at the site of the explosion and fire. Radiation spread over a large portion of the Soviet Union and Europe. Many scientists predict that the delayed effect of the radiation will cause cancer-related deaths for many years. Though the plant is back in operation, some areas around it may never be habitable.

The biggest problem with nuclear power may be a social problem. Some people are not willing to risk the dangers of nuclear accident. Nuclear power plants have been operating successfully for several years, but still they are not generally accepted. People still are afraid of the radiation hazards. This social problem must be overcome before nuclear power generation can be a complete success.

In the meantime, we continue with our energy-intensive way of living. We make more devices that need electrical energy. Hair dryers, electric can openers, and other appliances in the home require electricity. How will we generate this electricity in the future?

Communication

As one watches the evening news on television, the weather reporter shows satellite pictures of the cloud formations over the United States. Have you ever thought how much technology is involved in this very sophisticated moment of communication? Communication technology has had great effect on mankind. As with transportation, long ago the speed of communication was no faster than the speed of a fast horse. The Pony Express was the fastest communication method for many years.

This began to change with the invention of the telegraph. Through this new communication device people could communicate over long distances. This could be done in a matter of minutes. A message could be tapped one letter at a time using Morse Code. A reply message could be returned in the same way. The radio or *wireless* was developed and the time needed for communication was shortened. Messages could be sent over

The SBS telecommunications satellite is shown undergoing a pre-launch test.

long distances, almost instantly. With two-way radios people could talk over long distances the same as if talking face-to-face. The invention of the television added pictures to this communication system.

The development of communication satellites has extended radio and television reception to all parts of the earth. Radio and television signals can travel only so far. Mountains block television signals. People in deep valleys or remote areas have difficulty receiving the television signal. A satellite is put into orbit around the earth. Television signals are bounced off the satellite. The signals then can be beamed into these remote areas. More than two-thirds of the communities in Alaska cannot be reached by road or rail. Television communication by satellite can provide a vital link with the rest of the earth.

Communications satellites are controlled from this tracking station.

As one can see, one of the most important consequences of communication technology has been the improvement of communication among people. People in remote areas can get medical help through television. People in different parts of the world can learn about each other. They can communicate quickly and resolve misunderstandings. Understanding may prevent a small problem from growing to a big one. A telephone "hot line" exists between Washington, D.C. and Moscow for this purpose.

Typesetting and editing in modern newspaper plants is done on computers. This is an example of people talking to machines. The typesetters type into the computer terminal. They can see on the screen what they typed and can correct errors. When the copy is correct, they punch a button and the copy is put into the computer. When all copy is written, it can be printed on photographic paper. This is then cut and pasted to be photographed and printed by the offset process.

One consequence of this new automated process is the elimination of skilled typesetters. The job of typesetting is not

All the information needed to produce a page of type has been punched onto the tape.

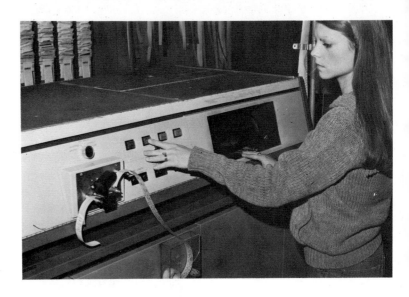

needed in this type of newspaper printing. These people must be retrained to operate the new equipment, or they must do some other work.

Why use this new process? It is faster and less expensive. It makes it possible to print a newspaper in less time. With some of these systems it is possible for a reporter to type a story directly into the computer. At the push of a button, the story is set into type immediately. A reporter at a big news event could connect a portable unit into a telephone line to the newspaper. The story could be written at the rate of 200 characters a second on the keyboard of the unit. The reporter now can write, edit, and have the copy in type without anyone handling a single piece of paper.

Manufacturing

Many of the consequences of manufacturing technology are well known. Mass-production techniques have caused manufacturing to be dehumanized. Production line workers must work hour-after-hour doing monotonous work. They are forced to work at the speed of the line. The conveyor moves. The work is repeated over and over again. A simple task may be done hundreds of times a day.

In some plants production line workers have revolted. They deliberately sabotaged the product. Some workers have walked off the job or unions have gone on strike for better working conditions. In some manufacturing plants the management has attempted to solve the problem. The workers have been reorganized. Work teams have been organized. Each team is responsible for building the entire product from start to finish. The workers are then able to take more pride in their work because they can see the product they are making.

Increased automation has caused great changes in jobs in manufacturing. The introduction of robots has eliminated the need for many production line workers. Some of these workers are retrained. Many may become unemployed. This has impact on both the workers and the communities where they live. As robotics is used more in industry this problem will increase. Will the many benefits of automation offset the negative effects?

There have been many consequences in manufacturing besides the organization of labor. Through manufacturing technology we now enjoy many new materials. Many of these have come from space research. For example, super-strong but light-weight graphite materials were developed for satellites. These same materials now are being used in sporting equipment. Racing bicycles can be made super-lightweight. Fishing rods and golf clubs are made of these materials also.

Foods developed for use in space have found their way onto the market. Highly nutritious foods that could be packaged with minimum weight and volume were needed in spacecraft. These foods now are being used in food packages in programs to feed the elderly or undernourished. They are used by backpackers, campers, and hunters.

Modern manufacturing has made many new products available. When Henry Ford began mass producing his first automobiles, he wanted to make an auto that could be sold at a price many could afford. By reducing his manufacturing costs he could sell the vehicle for less. High volume manufacturing can result in lower prices for consumers. The first televisions on the market were quite expensive. Few were produced until they were further developed and more companies were producing them. Today a small color television can be purchased for less than the first black and white sets cost.

Manufacturing has social consequences. We know that high-volume production requires high-volume consumption. If the automobile industry is going to produce millions of automobiles, then the consumers must buy millions of automobiles. This can cause problems. Advertising is used to convince people to buy the product. Sometimes the advertising convinces people to buy things they cannot afford. People may go into debt. Easy credit allows people to buy things without enough money. They promise to pay over a period of time. This is often a temptation to buy things that can't be afforded.

Construction

Construction technology has helped make our life more pleasant. We have comfortable houses in which to live. They protect us from the weather. Highways make transportation easier. We are able to travel nearly any place in the United States on a smooth highway. The interstate highway system provides wide, smooth roads between the major population centers. Modern construction techniques are used to build overpasses, bridges, tunnels, and interchanges to keep a smooth flow of traffic.

But, like the other technological systems, the consequences of construction technology are not all good. There are negative consequences from construction. Modern highways require space for the way. When the highway is built through a city, people must be moved. In many cases the life style of these people is

Sometimes buildings must be torn down and people relocated to make room for new construction.

changed. They move to a new neighborhood. They must make new friends and learn a new way of life.

A new bridge is built across a river in a city. Some people must be moved to make room for the bridge. Traffic patterns change as a result of the new route across the river. Streets that once were quiet residential streets are now major traffic routes. People who once lived on a quiet street now have to contend with heavy traffic. The quiet neighborhood changes to a noisy, busy area.

Construction can have impact on the environment. When a new highway is cut through the land, ugly scars often remain. Sometimes stone for the roadway is quarried from areas near the new road. To get the stone from a quarry farther away from the roadway would cost more. But, when the roadway is finished, the large holes in the earth remain. They are not filled or smoothed over because this costs money. They look bad and are dangerous when they fill with water.

Suggestions for Further Study

Activities

To help you better understand the consequences of technology you can become more aware of the happenings around you. Can you find examples of the counterforces listed in this chapter? Can you find ways that nature controls materials supply? How do people gain an advantage over nature? Find other examples of the counterforces.

Study the newspapers for reports of ways technology helps people. Can you also find ways it is harmful? Can you suggest ways some of these harmful consequences can be eliminated?

Questions

1. How does the use of technology help people gain advantage over nature?
2. List several ways that people create their own environment. Can you give examples from your home, your school and your community?
3. How have technological advances changed the way you use your leisure time?
4. What has happened in recent weeks that are examples of the positive and negative effects of technology?

Key Words

consequence leisure
counterforce expression
environment future
authority limitation
culture

Consequences of Technology

Concept: *Consequences*

Every technological development affects people in some way. Sometimes it is in a good way, sometimes bad. Once we understand how technology affects us, we can learn to live with technology. We must understand how we create and use technology. We must understand what prices we must pay for the benefits of technological developments.

You can learn more about the consequences of technology by reading the daily newspapers and weekly news magazines. Begin a scrapbook of articles about technology. A three ring binder will make a good scrapbook. Paste the clippings onto sheets of notebook paper.

Organize your scrapbook using the eight faces of technology presented in chapters one and ten of the textbook: materials, nature, environment, culture, leisure, change in humans, expression, work.

The several technological systems presented in the textbook could be used as another organizing scheme: energy and power, manufacturing, communication, transportation, construction.

Find examples of the ways technology affects the way people work, or the way they spend leisure time. How about environmental effects? How has technology contributed to such problems as "acid rain?" How can it help solve these problems?

What about the new composite materials being developed for high performance aircraft? Have you seen articles about attempts to make automobile engines from ceramic materials? These are examples of new materials affecting our technological world.

How have technological developments affected nature? What examples can you find of humans learning to live and work within the limits of nature? Practically everything we do is controlled by natural limitations.

What examples can you find of our culture being affected by technology? Have technological developments affected music, art or the theatre? Our cultural heritage is one of those things that make us the kind of people we are. It is the things that make us different from other cultures.

Look for ways people have changed the environment. What about the air conditioning of homes and automobiles? The refrigerant in some air conditioners is suspected to have an effect on the ozone layer surrounding the earth. What could be the effect of this on human life?

Spacecraft and space stations now being designed will have their own environments. In the space stations heat, light, and even the air breathed by the space colonists will have to be provided. How many environmental elements will have to be provided for space colonists?

What effect has technology had on work in your community? Are any industries using robots to do work? Where are computers being used?

These are just a few examples of the types of information you should be able to find in newspapers. As your scrapbook grows you should begin to develop a good understanding of the consequences of technology.

Your teacher may want to select some clippings and put them on a bulletin board. In this way you will be able to share and compare your ideas with your friends.

Glossary

Chapter 1 — What is Technology?

communication The process of exchanging or transmitting information.
computer A programmable machine capable of accepting, processing and outputting information.
machine A device that transmits forces, motion or energy.
production The process of making goods.
robot A mechanical device that can be programmed to move materials and perform specific tasks repeatedly.
synthetics Artificial materials, not human-made.
technology Applying the principles of science to achieve a practical purpose.
tools Human-made devices that extend human capabilities.
transportation The process of moving people and objects from one place to another.

Chapter 2 — Processing

bauxite A natural ore from which aluminum is extracted.
cable tool drilling Drilling into the earth by raising a bit attached to a cable, letting it fall repeatedly, punching a hole.
combining Putting together two or more materials.
conditioning Changing the characteristics of a material. For example, conditioning occurs when a metal is hardened by heating and cooling.
contour mining Mining by cutting away contours along hillsides.
dredging Deepening a waterway by removing sediments from the channel.
drift mines Mining process in which ore is removed from hillsides with the vein exposed.

347

extraction Removal or separation of metal from an ore.
farming The raising of livestock or crops.
grouping or classifying Organizing or arranging in some orderly way.
harvesting The process of gathering a crop.
mining Processes of removing solid materials from the earth.
nonrenewable resources Natural resources which cannot be replaced once they are depleted.
open pit mining Mining by removing earth from on top of the ore then digging out the ore. A large hole, or pit, results.
petroleum An oily, flammable bituminous liquid.
placer mining Mining by washing out materials with water.
postprocessing Actions or operations taken after the processing of a material.
preprocessing Actions or operations taken prior to the processing of a material.
processing Applying a series of actions or operations to materials to change them in some way.
pulp pulverized material mixed with water. Wood pulp is used in making paper.
quarry An open pit usually for the removal of stone.
raw materials Materials in a natural or unrefined state.
recycling The returning of a material to its original state so it can be reused.
reducing in size Making a material into smaller pieces.
refining Separating materials.
renewable resources Natural resources which can be replaced as they are used.
rotary drilling Drilling into the earth using a rotating cutter bit.
separating Removal of one material from another. Refining is a separating process.
shaft mines Mines using shafts or tunnels into the earth to reach the ore.
smelting Separating metal from ore by melting.
solid waste Refuse in a solid state, not liquid.
strip mining Mining by cutting away the surface of the earth in strips to get to materials below.

Chapter 3 — Energy and Power

conversion The changing from one form to another.
energy The capacity for doing work.
energy intensive Relating to the use of large amounts of energy.
fossil fuels Fuels formed by the deposit of fossils. Coal and petroleum are fossil fuels.
generator A machine which creates electrical energy.
geothermal Heat generated within the center of the earth.
helical gear A gear with the teeth in the form of a spiral or helix.
hydraulics Power transmission through fluid-filled lines.
hydroelectric Relating to the process of generating electricity from falling water. The water turns turbines to provide power to generators.

hydronic   Relating to the heating or cooling using the transfer of heat by a circulating fluid.
insulator   A material used to block the flow of heat or electrical energy.
internal combustion engine   An engine in which combustion occurs inside the engine, rather than in a separate combustion chamber.
link belt   A belt made of small segments fastened together.
nuclear energy   Energy derived from the atomic nucleus of uranium and other materials.
overshot wheel   A waterwheel turned by the flow of water over the wheel.
petroleum   An oily, flammable bituminous liquid.
photovoltaic   Relating to the process of converting radiant energy into electrical energy.
pneumatics   Power transmitted through air-filled lines.
power   The rate at which work is done.
rocket   An engine powered by the reaction to the exhausting of burning fuel through a nozzle.
roller chain   A chain made up of a series of rollers fastened together.
rotary engine   An engine in which a revolving rotor is used instead of pistons to create the combustion chamber.
solar cell   A device used to convert sunlight into electrical energy. A photovoltaic cell.
solar energy   Energy from the sun.
tidal energy   Energy generated by the movement of the ocean tides.
transformer   A device used to change voltage in an electrical circuit.
transportation   The process of moving people, products or materials from one place to another.
turbine   A series of curved vanes over which gases or liquids flow, causing the shaft of the turbine to rotate.
turbofan   A jet engine with a fan blade connected to the turbine shaft.
turbojet   An airplane powered by turbine engines.
turboprop   A turbojet engine with a propeller attached to the shaft of the turbine.
undershot wheel   A waterwheel turned by the flow of water beneath the wheel.
v-belt   A belt made to fit into a V-shaped pulley.

Chapter 4 — Manufacturing

advertising   Telling the customer about a product with the hope that it will be purchased.
assembly   The putting together of parts into a complete machine or unit of manufacture.
assembly line   An arrangement of machines and workers so work passes from operation to operation in line until the product is finished.
consumer   Person who purchases a product.
design   To plan an object for a specific function.
distribution   Moving or delivering a product from factory or warehouse to the customer.

feasibility study A study to determine if something can be done or done efficiently.
finances The funds used to defray the costs of an operation.
flow chart A chart showing the movement of materials or processes through a planned sequence of events.
human resources The talents and skills of the workers.
Industrial Revolution A period in history when a shift occurred from mainly agricultural economy to an industrial economy. This process began in England around 1760.
interchangeable parts Parts made to tolerances close enough so that any part can be used in place of any other.
just-in-time manufacturing The organization of production where needed parts arrive at the factory and assembly line when needed rather than being stored in a warehouse until needed.
management The people responsible for organizing and running an enterprise.
manufacturing The application of technology to produce products.
marketing The process of moving goods from producer to consumer.
natural resources Resources found in nature, e.g., raw materials and energy.
plant layout Organization of machines in a factory.
production The act of producing a product.
quality control The testing or checking of products or steps in manufacturing to assure a high quality product.
research and development The process of discovering basic facts or principles and the uses of these in practical operations.
retail The sale of goods directly to the consumer.
specialization of labor Reduction of work to one small act or job.
stockholders Persons who purchase stock or shares in an enterprise to help finance its operation.
testing The process of checking to see if a product or process meets standards.
wholesale The sale of goods for ultimate resale.

Chapter 5 — Construction

aluminum A bluish, silver-white metal, most abundant metal in the earth's crust. Extracted from bauxite ore.
architect A person who designs buildings.
asphalt A composition used for pavements.
bid An estimate of the cost of doing work.
canal A human-made waterway used for transporting materials in boats or ships.
carpenter A worker who builds structures out of wood.
concrete A hardened material made by mixing portland cement with an aggregate (such as sand or gravel) and water.
foundation The bottom-most support upon which a structure is erected.
inspector A person who checks for quality of workmanship.
insulation Materials used to block the flow of heat or cold.

management The people responsible for organizing and running an enterprise.
pyramid An ancient structure especially in Egypt, typically with a square ground plan, and four triangular sides rising to a peak.
slope Slant or incline, a roof or side of a hill for example.
solar heat Heat derived directly from the sun.
survey The taking of measurements to determine the location, slope and size of a track of land.
title A legal document stating ownership.
truss An assembly of beams to form a rigid framework.

Chapter 6 — Transportation

aileron Moveable surfaces on the wing of an airplane used to control roll around the longitudinal axis.
barge A flat-bottomed boat that is towed, not self-propelled.
controls Devices to start, stop or guide machines.
conveyor A mechanical apparatus for moving products or materials from place to place.
dry dock A dock that can be emptied of water for construction or repair of a ship.
escalator A moving stairway used to transport people up and down between floors of a building.
fuselage The body or main structure of an airplane.
general aviation Non-commercial aviation including sport flying.
gondola A train car with an open top and usually a flat bottom.
horizontal stabilizer Non-moving, horizontal surface on the tail of an airplane.
navigation The act of getting ships, aircraft or spacecraft from place to place along a planned route.
pipeline A system of pipes, pumps and controls for moving liquids and gases.
power plant A building where energy is produced.
rudder Vertical control surface on the rear of an aircraft or ship.
shipway The ways or base upon which a ship is built.
tanker A vehicle designed to carry liquids.
tramp freighter A cargo ship that does not operate on a fixed schedule.
transportation The process of moving people, products or materials from place to place.

Chapter 7 — Communication

broadcasting The transmission of a message to a large audience.
camera A light-tight box fitted with lens and shutter for recording images on light-sensitive paper.
channel or carrier The medium over which information is carried.
communication The process of exchanging or transmitting information.
director A person who supervises the action, lighting, and music for a media event.

f-stop A marking on a camera indicating the size of opening through which light passes. An indication of the amount of light reaching the film in a camera.
feedback Return of part of a message in a communication system.
halftone A picture made up of small dots of various sizes to form an image.
lens Transparent object with at least one curved surface. Focuses the light onto the film in a camera.
lithography A printing process based on the theory that water and oil do not mix. Offset printing is a modern use of the lithographic process.
microwave Electromagnetic wave of extremely high frequency.
noise Interference with the clear transmission of a message.
offset printing Lithography; plane-surface printing.
photography The process of producing images on a light-sensitive material.
producer A person who supervises or finances the production of a media event.
receiver The portion of a communication system that converts the transmitted message into visible or audible form.
script Written text of a media event.
shutter A mechanical device in a camera that regulates the length of time light is exposed to the film.
source of information The origin or generation of information to be put into a communication system.
telecommunication Communication through electronic means.

Chapter 8 — Automation
automatic control Control done by machine without human action.
automation Operation of machines without direct human control; machines controlling other machines.
axis A degree of freedom or a basic motion allowed by a mechanism.
bimetallic Two metals usually joined in some way.
conveyor A mechanical apparatus for moving products or materials from place to place.
degree of freedom The different ways a robot's limb can move — up, down, left, right, rotate, extend, etc.
detector A device for sensing the presence of electromagnetic waves.
production The process of making goods.
regulate To control.
relay A switching device for remote or automatic control.
self-correcting Automatic feature of a machine which adjusts automatically to changing conditions.
solenoid A coil of wire through which a metal core passes. When the coil is magnetized, the core moves.
thermostat Automatic device for regulating temperature.
work-envelope The region which can be reached by a robotic arm. This is limited by the length of the arm, joint arrangement, range of motion of the joints and the load being carried.

Chapter 9 — Computers

abacus An ancient device for calculating.

BASIC Beginners All-Purpose Symbolic Instruction Code: A simple language for programming computers.

CAD Computer Assisted Drafting: the production of drawings through the use of a computer.

calculator A device used to do mathematical calculations.

computer A programmable machine capable of accepting, processing and outputting data.

core memory Computer memory in the computer. This is internal memory as opposed to external memory such as disks or tapes.

CPU Central Processing Unit: the electronic circuitry in a computer that interprets and executes instructions. This is the "brain" of the computer, the main control unit.

daisy wheel printer A printer with a round print element resembling a flower or "daisy." The printed image is formed when a hammer strikes one of the letters on the element driving it against an inked ribbon and onto the paper.

data base Computer programming used to control large amounts of data such as lists, research information, etc.

dot matrix printer A printer that uses dots to form the printed image.

electronic tube An electronic device to perform amplifying, oscillating, rectifying or switching functions. It is made in a glass or metal enclosure with no air.

floppy disk A thin, round disk of magnetized plastic material used to store computer data.

graphics tablet A computer input device used for putting line drawings and other graphics into a computer. It is made with a wire grid embedded into a base. A stylus is used to touch the base, causing the wires to touch, sending a signal to the computer.

ink jet printer A computer printer using dots of liquid ink to form the image. The ink is sprayed from tiny "ink jets" or nozzles onto the paper.

integrated circuit An electronic circuit with all components formed on a piece of semiconductor material. This material is usually silicon.

joy stick A computer input device that looks like a stick on a box. The stick is hand-held and moved up-down and left-right to move the cursor on the computer monitor.

keyboard A computer input device similar to a typewriter keyboard.

letter quality Refers to a quality of computer printer output that is the same quality as printed material or high quality typed letters.

light pen A light sensing computer input device. It resembles a pencil. It is used by touching the screen of the computer monitor. This conveys information to the computer.

mainframe computer The largest computers usually capable of handling several users at the same time. Mainframe computers operate much faster than other, smaller computers.

memory Electronic circuitry in a computer used to hold programs and data.

microcomputer A small computer, usually a desktop or portable size; often called a home computer or personal computer.

mouse A hand-held computer input device used to move the cursor around on the computer monitor. The mouse is rolled around on a flat surface. The ball or wheels inside the mouse control the cursor.

pixel One of the thousands of "dots" or picture elements on the screen of a computer monitor.

RAM Random Access Memory: usually referred to a read-and-write-memory. A form of temporary memory in the computer. The user can add to RAM. All information in RAM is lost when the computer is turned off.

relay An electromagnetic device used for automatic or remote control, usually of other devices.

ROM Read-Only Memory: permanent memory in the computer that can not be altered by the user. It can only be used as programmed by the manufacturer.

software Computer program or instructions. In contrast, computer hardware refers to the machines such as the computer, keyboard, monitor, etc.

spread sheet Computer programming used for accounting and bookkeeping.

terminal An input device on a computer usually composed of a keyboard and a monitor.

transistor An electronic device which replaced the electronic tube for many uses.

word processing Computer programming used in writing.

Chapter 10 — Consequences of Technology

authority The power to influence or command; a person in power; an expert.

consequence Some thing or act that is caused by another thing or act.

counterforce Acting in opposition to a force.

culture The patterns of human behavior transmitted to succeeding generations.

environment The climatic surroundings.

expression A significant work or phrase.

future What is going to happen; time yet to come.

leisure Time away from work.

limitation Something that limits or restricts.

Readings

Chapter 1 — What Is Technology?

Many books and magazines deal with technology. Some are trade magazines that are written for people working in various fields. Others are of a more general nature and are written for general interest. Look in your school and community libraries. Small libraries usually do not have many of the trade publications because they are not of general interest. Look for the books and magazines with a more general emphasis.

The following books, written and illustrated by Eric Sloane, are excellent records of technology in early America. Each is written in a readable style and illustrated with the author's own sketches.

ABC Book of Early Americana
An Age of Barns
American Barns and Covered Bridges
American Yesterday
The Cracker Barrel
Diary of an Early American Boy: Noah Blake—1805
Eric Sloane's Almanac
A Museum of Early American Tools
Our Vanishing Landscape
Return to Taos: A Sketchbook of Roadside Americana
A Reverence for Wood
The Seasons of America Past

Similar books have been written and illustrated by Edwin Tunis and other authors. These should be interesting as you study technology. They also are written about early stages of technology.

The *Foxfire Book, Foxfire II, Foxfire III,* and *Foxfire IV,* are records of early American technology. These publications are collections of articles written by high school students.

As you continue your study of technology, you will discover many books and magazines that deal with technology, at least in part. *Popular Mechanics, Popular Science,* and *Mechanics Illustrated* will be particularly interesting. Each has many features dealing with modern technology. Your school and community libraries should have these magazines.

Many others are written for specific areas of technology. These will be mentioned in the later chapters of this book.

Your teacher or your school or community library will have a list of publications related to technology.

Chapter 2 — Processing Technology

Facts about Oil Washington, D.C.: American Petroleum Institute
Products of American Forests Washington, D.C.: U.S. Department of Agriculture
The Making of Steel. Washington, D.C.: Public Relations Department, American Iron and Steel Institute
The Story of Coal Bethlehem Steel Co.: Public Affairs Department
The Story of Petroleum. Houston, Texas: Shell Oil Company

Chapter 3 — Enery and Power

Bohn, Ralph C. and McDonald, Angus J., *Power: Mechanics of Energy Control* Bloomington, Ill.: McKnight & McKnight Co., 1970.
Boy Scouts of America, *Energy.* Merit Badge Pamphlet.
Duffy, Joseph, *Power: Prime Mover of Technology.* McKnight & McKnight, 1964.
McKenzie, Bruce A. and Zachariah, Gerald L., *Understanding and Using Electricity.* Danville, Ill.: The Interstate Printers & Publishers, 1975.
Stewart, Harry, L. and Storer, John M., *Fluid Power.* Indianapolis: Howard Sams & Co., 1968.
Suess, Alan, et al., *Introduction to Power Systems,* Beverly Hills, Calif.: Bruce, 1976.
Yadon, James N., *Energy and Transportation: Small Engines.* Englewood Cliffs, N.J.: Prentice-Hall, Inc., 1976.

Magazines & Pamphlets

Eaton, William W., *Energy Technology,* Energy Research and Development Administration, Washington, D.C. 1975.
Jet Know-How, Pratt & Whitney Aircraft Co., East Hartford, Conn.
Popular Science
Popular Mechanics

Chapter 4 — Manufacturing

There are many ways you can become active with manufacturing. The activity manual with this book suggests several ways. Most can be done in your school. Some can be done at home or in your community. Hobbies are a good way to learn about manufacturing.

Chapter 5 — Construction

Feirer, John L. and Hutchings, Gilbert R. *Carpentry and Building Construction.* Peoria, Ill.: Chas, A. Bennett Co., Inc. 1976.

Lux, Donald G. and Ray, Willis E. *The World of Construction.* Bloomington, Ill.: McKnight Pub. Co., 1970.

——— *Readings in World of Construction.* Bloomington, Ill.: McKnight Pub. Co., 1977.

Smith, R. C. *Materials of Construction.* New York. McGraw-Hill Book Co., 1966.

Spence, William P. *Architecture: Design, Engineering, Drawing.* Bloomington, Ill.: McKnight Pub. Co., 1979.

Chapter 6 — Transportation

Angelucci, Enzo. *Airplanes From the Dawn of Flight to the Present Day.* New York: McGraw Hill, 1973.

Baker, Elijah, III. *Introduction to Steel Shipbuilding.* New York: McGraw Hill, 1953.

Banbury, Philip. *Man and the Sea.* London: Adlard Coles Ltd., 1975.

Bingham, Truman C. and Roberts, Merrill T. *Transportation: Principles and Problems,* New York: McGraw Hill, 1952.

Branch, Alan E. *The Elements of Shipping.* London: Chapman and Hall, 1975.

Danforth, Paul M., *Transportation: Managing Man on the Move.* New York: Doubleday & Co., 1970.

Faulks, R. W., *Principles of Transport.* London: Ian Allan Ltd., 1973.

Firestone, Harvey S. *Man on the Move.* New York: G. P. Putnam's Sons, 1967.

Fishbaugh, Charles. *From Paddle Wheels to Propellers.* Indiana Historical Society, 1970.

Gibbs-Smith, C. H. *Flight Through the Ages.* New York: Thomas Y. Crowell Co., 1974.

Gunston, Bill. *Transport: Problems and Prospects.* London: Thomes and Hudson Ltd., 1972.

Jane's, *All the World's Aircraft*. New York: Arco, 1972.

Jane's, *World Railways*. New York: Arco, 1977.

Munro-Smith, R. *Merchant Ship Types*. London: Marine Media Management, Ltd., 1975.

Sampson, Roy J. and Farris, Martin T. *Domestic Transportation,* Boston: Houghton Mifflin, 1975.

Smith, Norman F. *Wings of Feathers, Wings of Flame: The Science and Technology of Aviation*. Boston: Little, Brown Co., 1972.

VonBraun, Wernher, and Ordway, Frederick I., III, *History of Rocketry and Space Travel*. New York: Thomas Y. Crowell Co., 1975.

Ridley, Anthony, *An Illustrated History of Transportation,* Scranton, Pa.: The John Day Co., 1969.

Chapter 7 — Communication

Becker, Joseph. *The First Book of Information Science*. Washington, D.C.: U.S. Energy Research and Development Administration, Office of Public Affairs, 1973.

Corliss, William R. *Computers*. Washington, D.C.: U.S. Energy Research and Development Administration, Office of Public Affairs, 1973.

Brix, V.H. *You Are A Computer*. New York: Emerson Books, Inc., 1970.

Meadows, Charles T. *Sounds and Signals—How We Communicate*. Philadelphia: The Westminster Press, 1975.

The Way Things Work Book of the Computer. New York: Simon & Schuster, 1974.

Chapter 8 — Automation

The Way Things Work Book of the Computer. New York: Simon and Schuster, 1974.

Hancock, Ralph (ed.), *How it Works: The Illustrated Encyclopedia of Science and Technology*. New York: Marshall Cavendish.

McGraw Hill Encyclopedia of Science and Technology, Vol. 1.

Chapter 9 — Computers in Technology

Darling, David. *Computer Graphics*. Minneapolis, MN: Dillon, 1987.

Hellman, Hal. *Computer Basics*. Englewood, NJ: Prentice-Hall, Inc., 1983.

Laron, Carl. *Computer Software Basics*. Englewood, NJ: Prentice-Hall, Inc., 1985.

Markle, Sandra. *Computer Tutor: An Introduction to Computers*. Santa Barbara, CA: The Learning Works, 1981.

Naiman, Arthur, ed. *Computer Dictionary for Beginners*. New York, NY: Ballantine Books, Inc., 1983.

Computer Programming in BASIC for Everyone. Fort Worth, TX: Radio Shack.

Reffin & Smith. *Computer Programming.* Tulsa, OK: Usborne-Hayes, 1982.

Relf, Patricia A. *Computer Thinking.* New York, NY: Scholastic, Inc., 1984.

Smith, David F. *Computer Dictionary for Kids & Other Beginners.* New York, NY: Ballantine Books, Inc., 1984.

Stockley, C. & L. Watts. *Computer Jargon.* Tulsa, OK: EDC Publishing, 1983.

Sullivan, George. *Computer Kids.* New York, NY: Dodd, Mead & Co., 1984.

Tatchell. *Computer Graphics.* Tulsa, OK: EDC Publishing, 1984.

Computer Languages. Alexandria, VA: Time-Life Books, 1986.

Waters. *Computer Fun.* Tulsa, OK: EDC Publishing, 1984.

Wold, Allen L. *Computer Science.* Danbury, CT: Watts, Franklin, Inc., 1984.

Chapter 10 — Consequences of Technology

Bova, Ben. *Man Changes the Weather.* Reading, Mass.: Addison-Wesley Publishing Co., 1973.

Lefhorwitz, R.J., *Fuel for Today and Tomorrow*, New York: Parent's Magazine Press, 1974.

Marshall, James. *Going to Waste.* New York: Coward, McCann & Geoghegan, 1972.

Wise, William. *Fresh, Canned, and Frozen: Food From Past to Future.* New York: Parent's Magazine Press, 1971.

Woodburn, John H. *The Whole Earth Energy Crisis.* New York: G.P. Putnam's Sons, 1973.

Index

Advertising, 131–34
Air taxi, 190
Air travel, 151,
Airplane, 151–52, 184, 186, 186–92,
    220, 232
  controls, 227
  parts of, 210–11
Airport, 151
Airways, 231
Aldrin, Edwin, W., 188
Alternating current, 69
Aluminum, 48–9, 213
Amplification, 265
Amplitude modulation, 256–58
Apollo-Soyuz, 188
Apprenticeship, 174
Architect, 163
Armstrong, Neil A., 188
Assembly, 122, 127
Assembly line, 9, 111–12, 208
Astronauts, 7, 12–13, 29, 59
Atomic energy, 116
Automation, 275–96
  components of, 275
  definition, 275
  effects of, 289
Automobile, 10, 203, 225–26

Barges, 200
Bauxite, 48
Bell, Alexander Graham, 6–7, 248,
    249
Belt drive, 91
Bicycle, 224

Brady, Matthew, 243, 234
Brakes, 94–95
Bricks, 157
Bridges, 149, 229
Building, 145–46, 164–73

Caboose, 208
Camera, 243–45
Capek, Karel, 284
Careers, 50–52, 123–25,   134–37,
    173–76, 232–34, 269–70
Cargo ships, 199–20
Carpenters, 165
Chernobyl, 339
Change 331–32
Citizen's Band Radio, 331
Civil aviation, 232
Coal, 30, 33–34, 37, 216
Coal mines, 149
Collectors, 76
Communication
  definition, 239
  history, 240–41
  impact of, 265
  systems, 7
  visual, 241
Compact Disc, 266
Complexity of technology, 19
Computers, 19, 118–19, 268–69,
    277, 283, 287
  BASIC, 317
  Chips, 308
  CPU, 307
  definition, 306

361

    development of, 297-99
    floppy disks, 310
    hardware, 308
    mouse, 310
    RAM, 307
    ROM, 307
    word processing software, 315
Concrete, 158
Conservation, 102
Contour maps, 161
Contractor, 173
Construction, 344-45
Consumer, 138-39
Controls, 211, 225-29, 279-80
Control tower, 152
Conveyor, 192-93, 343
Conveyor belt, 282
Copper, 15
Culture, 331
Cybernetics, 19

Daguerre, Louis, 243
DeForest, Lee, 248
Design, 117
Diesel engine, 81-82, 219
Distillation, 41
Dividends, 113
Double bottom assembly, 202
Drag, 213
Drilling, 34-35
Duryea, Charles E., 216
Dwellings, 147-48

Eastman, George, 117, 243
Education, 135-37
Electrician, 170
Electricity, 68, 70, 88-90, 216, 248
Electrical distribution, 88
Electromagnetic waves, 256
Energy, 5
    conversion, 68
    definition, 57
    history, 57-58, 63
    resources, 116
    systems, 5
    uses, 95-98
Energy intensive, 101, 337, 339
Engines, 78-87
Environment, 12-13, 116, 329
Escalator, 196
Evans, Oliver, 79
Extraction, 31

Factory, 5-6
Faraday, Michael, 68
Feasibility study, 122
Federal Aviation Administration (FAA), 151
Feedback, 283
Fiber optics, 249
Film, 244
First tools, 1-2, 154-155
Flight, 227-228
Flow charting, 123-124
Fluid power, 93-95
Flying, 189-192, 212-213
Food, 25, 147, 343
Food and Drug Administration (FDA), 139
Ford, Henry, 206
Fossil fuels, 62-63
Franklin, Benjamin, 248
Freight trains, 207-08
Frequency modulation, 257-57
Fuel, 3-4, 62-63

Gasoline, 42-43
Gasoline engine, 80-81
Gagarin, Yuri, 188
Gears, 93
Generator, 68
Geothermal energy, 66-67, 74-75, 338
General aviation, 184-85
General cargo ships, 198
Glass, 159
Gutenberg, Johann, 241-42

Hand tools, 5
Health, 9-10
Heat loss, 100
Hieroglyphics, 241
Highways, 229-230
Hobbies, 12, 21, 104, 270, 331
Human resources, 113-14
Hydraulic, 32, 93-95
Hydroelectric power, 62, 72-73

Industrial revolution, 6, 79, 111
Insecticides, 11
Insulation, 170
Insulators, 90
Interchangeable parts, 8, 112
Internal combustion engine, 80-88, 216

362

Jet engine, 84–87, 220–21
Jet propulsion, 84
Journeyman, 174

Keel, 202

Labor, 113
Land, 160–62
Land development, 160
Land, Edwin, 118
Langley, Samuel P., Jr., 184
LASH ship, 200
Leisure, 11–12, 104, 331
Letterpress printing, 247
Lindbergh, Charles, 187
Lithography, 242, 245
Locomotive, 207, 217
Lunar Rover, 7

Machines, 5, 104
Mchine tools, 79–80
Machining, 282
Magnetic field, 68, 90
Marconi, Guglielmo, 248
Market survey, 121, 130
Mass media, 261
Mass production, 342
Mass transportation, 337
Materials, 328
Materials handling, 122–23
Mechanical power transmission, 91–92
Mechanication, 277
Merchant marine, 232–33
Mining, 31–35
Modern materials, 4, 15
Modular construction, 159
Monorail, 231
Morse, Samuel F. B., 248
Moving way, 193, 196

Natural materials, 3
Natural resources, 115
Nature, 328
Navigation, 182
Newcomen, Thomas, 79
Non-renewable resources, 102
Nuclear energy, 63–64, 339
Nuclear power, 63, 70, 219, 339

Occupational Safety and Health Act OSHA, 129
Oil, 8, 25, 29, 35, 40, 42

On-the-job-training, 173
Organization chart, 115
Ore, 30–36

Panama Canal, 150
Paper, 43–47
Personnel, 113–15, 134
Petrochemicals, 42–43
Petroleum, 42, 98
Pewter, 39
Photography, 243–247
Photovoltaic, 76
Pipeline, 36, 193–94
Plant layout, 122
Plastics, 14, 121
Plumber, 166
Pollution, 10
Power (definition), 57
Power plants, 215–16, 218
Power transmission, 88–90
Printing, 241–42, 245–47
Processing, 27, 38–39, 283
Product development, 117
Product research, 119
Product survey, 121
Production, 18, 112, 122, 291

Quality control, 123, 128

Radio, 6, 209, 255–59, 264
Railroads, 149, 230
Recycling, 30, 103
Refining, 40–43
Research and development, 115
Resources, 3–4, 15–16, 18, 113–116
Renewable resources, 30
Roads, 115
Robot, 285, 294
Robotics, 284
Rockets, 87, 186–87, 221–23
Rubel, Ira, 242
Runway, 152

Safety, 129
Sales promotions, 131
Satellites, 13, 188, 252–53, 339
Sensing, 276, 278
Ships, 57, 218–19
  building, 201–03
  control, 226
Shipway, 203
Site work, 164

363

Skylab, 76
Solar energy, 60, 65–66, 76, 154, 338
Space craft, 12
Space Shuttle, 223
Space travel, 221–24
Specialization of labor, 111, 112
Standard of living, 17, 105
Steam engine, 5–6, 78–80, 182, 184–85, 216
Steel, 15, 26, 158
Steering, 225
Surveying, 160, 164
Systems, 327, 334
  highway, 229
  production, 8
Synthetics, 16

Tankers, 201, 209
Technology
  contexts, 18
  definition, 1–3
model of, 17–18, 327–328
Telecommunication, 249
Telephone, 6, 249, 341
Television, 6, 248, 259–261, 264, 340
Tidal energy, 66, 73, 338
Tools, 1, 125, 148, 154–55
Topographical survey, 161
Trains, 206–09, 216–18, 226
Transformer, 90
Transmitting energy, 88–91

Transportation, 18
  air, 184
  components of, 197
  development of, 182–86
  energy in, 100–01
  highway, 229
  land, 182
  space, 186–88
  systems, 7
  water, 183
Trevithick, Richard, 79
Trucking, 234
Trucks, 205
Turbine, 67–71, 217–21
Turbofan, 85, 87
Turbojet, 85, 221
Turboprop, 86, 87, 221

Underwriters Laboratory (UL), 138
Utilities, 169

Vehicles (transportation), 197–215

Water, 64, 66, 72–80
Waterway, 150, 232
Watt, James, 79
Weather, 13
Whitney, Eli, 111
Wind, 57, 65, 75–76
Work, 14, 15, 333–334
Work envelope, 286
Wright, Orville & Wilbur, 6, 151, 184

Picture Credits

Illustrations
S. Whitney Powell pp. 29, 34, 41, 69, 65, 68, 71, 71, 81, 85, 86, 87, 88, 89, 90, 93 top, 94, 149, 161, 165, 166 top, 167, 169, 194, 202 left, 210, 212 top, 213, 225, 227, 228, 243, 244, 246 top, 247 bottom, 256, 257, 260, 264, 280.
Linda Sturgis pp. 18, 62, 96, 101, 115, 124, 166 bottom, 328.

Photographs
© 1979 ABC, Inc. p. 269 right.
Alinari Editorial Photocolor Archives pp. 146 top left, 146 top right.
The Aluminum Association, Inc. pp. 48, 49, 50, 326.
American Iron and Steel Institute p. 35 bottom.
American Petroleum Institute p. 35 bottom.
American Trucking Associations, Inc. p. 37.
Association of American Railroads, pp. 7 bottom, 185, 217.
Atchison, Topeka and Santa Fe Railroad, p. 80.
A.T.&T. pp. 119, 135 right, 252.
Allen Bame pp. 58 top, 211 top.
Bethlehem Steel, p. 100.
B.F. Goodrich Company pp. 16, 51.
Bicycle Manufacturers Association of America p. 11.
Black Dot, Inc. Ed. Osech pp. 246 bottom, 247 top, 269 left, 342.
Reproduced with permission of the Coca-Cola Company p. 274.
Coca-Cola Foods Division, Houston, Texas pp. 129, 333.
Paul Cummings pp. 158, 160, 164 left, 164 right, 170 bottom.
Dayco Corporation's Rubber Products Company, Dayton, Ohio pp. 91, 92.
Deere and Company pp. 58 bottom, 59.

Dole p. 38.
E.I. du Pont de Nemours and Company pp. 110, 136.
Eaton Corporation p. 192.
Eli Lilly and Company pp. 1, 135.
The Energy Bank p. 171.
Florida Department of Citrus p. 195.
Ford Motor Company p. 8.
French Embassy p. 73.
French Government Tourist Office p. 146 bottom left.
General Motors Corporation pp. 127 top, 206.
Grumman Energy Systems p. 154.
Gulf Oil Corporation pp. 4, 338.
Hammermill Paper Company pp. 45 left, 46, 47 bottom.
H.P. Hood, Inc. p. 282.
IBM pp. 113, 128.
International Ladies' Garment Workers Union p. 113 left.
International Paper Company pp. 44 right, 47 top.
Mannheim Manufacturing and Belting Company p. 92 bottom.
Massachusetts Port Authority pp. 151, 199.
NASA pp. 12, 66, 221, 223.
National Association of Independent Schools p. 104.
Norfolk and Western Railroad pp. 208 (3 photos), 231 right, 231 bottom.
Oil and Gas Journal p. 40.
Pacific Gas and Electric Company pp. 67, 70.
Polaroid Corporation pp. 133, 139.
Pratt and Whitney Aircraft pp. 85, 86 (3 photos), 189, 220.
R.C.A. pp. 253 bottom, 341.
School Arts p. 75.
Schwinn Bicycle Company p. 93 bottom.
Texaco, Inc. O. Winston Link pp. 193, 200.
Transportation Systems Center, Cambridge, Massachusetts p. 336.
Union Pacific Railroad Museum Collection p. 14.
U.S. Air Force Photo p. 187.
U.S. Army Corps of Engineers pp. 43, 175.
U.S. Department of Housing and Urban Development pp. 146 bottom right, 153 left, 153 right, 159, 174, 344.
United States Steel pp. 30, 32.
Water and Power Resources Service, U.S. Department of the Interior p. 56.
Western Electric pp. 7, 8 top, 127 bottom, 252 top.
Western Wood Products Association pp. 155, 156, 157, 172.
Westinghouse p. 10.
White Motor Corporation p. 276.
WNAC-TV, Boston p. 262.
From *The World Book Encyclopedia*. Art Work by Tom Dunnington (© 1979 World Book-Childcraft International, Inc.) p. 242.